Rice Productivity and Food Security in India

CMA Publication No. 250

Poornima Varma

Rice Productivity and Food Security in India

A Study of the System of Rice Intensification

Springer

विद्याविनियोगाद्विकासः
IIM
AHMEDABAD

Poornima Varma
Centre for Management in Agriculture
Indian Institute of Management Ahmedabad
Ahmedabad
India

ISBN 978-981-10-3691-0 ISBN 978-981-10-3692-7 (eBook)
DOI 10.1007/978-981-10-3692-7

Library of Congress Control Number: 2017940606

Printed on acid-free paper

This Springer imprint is published by Springer Nature
The registered company is Springer Nature Singapore Pte Ltd.
The registered company address is: 152 Beach Road, #21-01/04 Gateway East, Singapore 189721, Singapore

Foreword

It gives me great pleasure to present to you this important work of the Centre for Management in Agriculture (CMA), Indian Institute of Management, Ahmedabad, India (IIMA). This is the 250th work in our series of published books and monographs since 1971. CMA has been actively engaged in research on important current topics and challenges in the management of the food, agriculture, agribusiness and rural sectors of the country and the world. The Centre is supported by the Ministry of Agriculture & Farmers Welfare, Government of India, and undertakes policy and evaluative research studies for the Ministry, and also conducts research on its own and for other national and international organisations. This spans policies and initiatives in topics such as the management of agricultural inputs, agro-processing, agri-food marketing, rural infrastructure, institutions, grass-roots innovations, appropriate technologies for agriculture, new technologies like GMOs, international agricultural trade and WTO issues, global competitiveness, commodity markets, food safety and quality including organic food, food supermarkets, food value chains and farmer collectives such as producer companies.

Food security continues to be a major challenge for a huge number of countries in the world, and within this, rice is the staple food for more than half of the world's population. Rice is the number one food staple in India, and being the second largest producer and consumer of rice in the world, India also substantially influences the global rice economy. There has been a downturn in the rice productivity growth in India in the recent years, and there is an urgent need for new breakthroughs in rice production technologies to ensure food security. The System of Rice Intensification (SRI) is a significant and promising new technology which not only boosts rice productivity substantially but is also more environment-friendly and sustainable than existing conventional technologies.

The present research examines the adoption of this promising technology by Indian farmers, studying the factors influencing adoption, the intensity of adoption, and the welfare gains possible through it in major rice producing states of India. It is based on primary data collected through a comprehensive household survey of 386 SRI adopters and non-adopters spread across southern, central and eastern regions of India spanning six districts in the states of Karnataka, Madhya Pradesh and Orissa. The results offer unique policy-relevant insights on the adoption of new technologies, the impact of SRI on improving rice productivity and household incomes, as well as the constraints the farmers face, and the need to address them with appropriate policy interventions.

I am sure the study will be found useful by policymakers, researchers, and those seeking to promote new technologies such as SRI, as well as others interested in agricultural policy, new technology adoption, food security, and the welfare of farmers.

June 2017

Vasant P. Gandhi
Chairperson
Centre for Management in Agriculture
Indian Institute of Management Ahmedabad
Ahmedabad, India

Acknowledgements

This study would not have been possible without the generous help and support of several individuals and institutions. I would like to express my sincere gratitude to the Ministry of Agriculture, Government of India, for financial support in undertaking this study. My special thanks to Prof. Vasant P. Gandhi, Chairperson of CMA, for his encouragement, support and valuable comments throughout the period of this study. My deepest gratitude also goes to Prof. Sukhpal Singh, Prof. Anil Gupta, Prof. Vijay Paul Sharma and Prof. Vaibhav Bhamoria for their constant encouragement and valuable comments.

I would like to extend my sincere gratitude to the Department of Agriculture, Hassan, especially assistant agricultural officer (AAO) and technical officers of various Taluks in Hassan for providing key information and data regarding the System of Rice Intensification (SRI) in Hassan district of Karnataka.

The study of SRI in Chikmagalur district of Karnataka would not have been possible without the help received from Shree Kshetra Dharmasthala Rural Development Project (SKDRDP). Staff members of SKDRDP provided generous help in primary data collection in various Taluks of Chikmagalur.

My special thanks to Mr. Lokesh for his immense contribution in the form of translation and data collection in Karnataka.

I am very grateful to the officers of the Department of Agriculture, Keonjhar, Orissa, for their support in conducting this study. My special thanks to assistant agricultural officer and field officer of various Taluks in Keonjhar for providing information regarding implementation of NFSM-SRI scheme in Keonjhar district.

I am thankful to PRADAN especially Mr. Anjan and his team in Keonjhar for providing immense local assistance in connecting with SRI and non-SRI farmers of the region. The same team extended their generous help in conducting field study in Mayurbhanj district of Orissa as well.

My study benefited a lot from the support of PRADAN in Sidhi district of Madhya Pradesh. I record my deepest gratitude to Mr. Saroj Barik and his team, PRADAN, for his support in collecting the data from Sidhi district of Madhya Pradesh.

My sincere gratitude also goes to the officers of the Department of Agriculture, Shahdol, especially the field officer Mr. Akhilesh Namdev. He has contributed a lot to this study by helping us in field survey in Shahdol.

I would like to extend my special acknowledgement to Ms. Kanak Khare for her invaluable contribution to this study. This study would not have materialised without her enthusiastic efforts in collecting field data and compilation. My special thanks also go to Mr. Akash Issar for his invaluable help and support in analysing the data. Ms. Sonali Kaur Bhatia deserves special acknowledgement for going through my final draft of this study and giving valuable comments and suggestions. I owe a lot to Peter for painstakingly going through this study and in editing.

Thanks are also due to Prof. C. Shambu Prasad, Prof. K. Palanisamy and Dr. Shouvik Chakraborty for their valuable comments and suggestions on the draft version of this study. Also, I would like to express my deepest gratitude to Prof. Norman Uphoff for enriching my study through the discussions we had at the final stage of this study.

At this juncture, I would like to also thank the CMA staff members for their generous administrative support in each and every stages of this study. Needless to mention while I owe a lot to the numerous individuals who contributed significantly to conduct this study, the data analysis and the views expressed in this report and any omissions or errors that remain in the report are entirely mine.

Poornima Varma

Contents

About the Author

Poornima Varma is an Assistant Professor at the Centre for Management in Agriculture (CMA) of the Indian Institute of Management Ahmedabad (IIMA). She holds a PhD in economics from the Centre for Economic Studies and Planning at Jawaharlal Nehru University, New Delhi. She was a recipient of a US Fulbright scholarship in 2006 and was subsequently a visiting research scholar at the Department of Economics of the University of Massachusetts Amherst, USA, during 2007 to 2008. Prior to joining the IIMA, she worked as an Assistant Professor at the Department of Policy Studies of TERI University, New Delhi. She has also worked with the Centre for WTO Studies (CWS) of the Indian Institute of Foreign Trade (IIFT) and the Indian Council for Research on International Economic Relations (ICRIER). Her work chiefly focuses on international trade, agriculture, environment and the adoption of agricultural technologies. She has extensively published in top peer-reviewed journals for agricultural and applied economics. She has also presented her papers in various national as well as international conferences including the European Trade Study Group conference in Vienna, Austria and the Agricultural and Applied Economics Association conference (AAEA) in Boston, USA.

Abbreviations

AAS	Academy of Agricultural Sciences
AFSC	American Friend's Service Committee
ATMA	Agriculture Technology Management Agency
AWD	Alternate wetting and drying
BMP	Best Management Practices
BPL	Below poverty line
BRLPS	Bihar Rural Livelihoods Promotion Society
CADA	Community Area Development Authority
CIIFAD	Cornell International Institute for Food, Agriculture and Development
CMP	Conditional mixed process
CRRI	Central Rice Research Institute
CSA	Climate smart agriculture
CSO	Civil society organizations
CSP	Community service provider
DH	Simultaneous equation
DPRK	Democratic People's Republic of Korea
DSR	Direct seeded rice
EU	European Union
FAO	Food and Agriculture Organization
FFS	Farmer's Field Schools
GDP	Gross domestic product
GNU	Gyeongsang National University
HYV	High yielding variety
ICDP	Integrated Cereal Development Program
ICT	Information and Communication Technology
IRRI	International Rice Research Institute
J&K	Jammu & Kashmir
MARDI	Malaysian Agricultural and Rural Development Institute
MML	Mixed multinomial logit
MSP	Market selling price

MVP	Multivariate probit
NABARD	National Bank for Agriculture and Rural Development
NE	North East
NFSM	National Food Security Mission
NGO	Non-governmental organization
NREGA	National Rural Employment Guarantee Act
OP	Ordered probit
PDS	Public distribution system
RKVY	Rashtriya Krishi Vikas Yojana
SCI	System of Crop Intensification
SFMI	System of Finger Millet Intensification
SICA	Sistema Intensivo de Cultivo Arrocero
SKDRDP	Shri Kshetra Dharmasthala Rural Development
SRI	System of Rice Intensification
SSI	Sustainable Sugar Initiative
STI	System of Teff Intensification
STRASA	Stress-Tolerant Rice for Africa and South Asia
SWI	System of Wheat Intensification
TC	Tissue culture
UKM	Universiti Kebangsaan Malaysia
WTO	World Trade Organization

List of Figures

List of Tables

Chapter 1
Introduction

1.1 Introduction

Rice is the staple food for more than half of the world's population and plays a pivotal role in food security of many countries. It constitutes nearly 26% in total cereal production and nearly 20% in total cereals trade (FAOSTAT 2014). Due to its importance in providing national food security, most Asian governments regard rice as a strategic commodity. Maintaining self-sufficiency in rice production and ensuring stability in rice prices have remained important political objectives in most Asian countries.

As the second largest producer and consumer, India plays an important role in the global rice economy. However, rice cultivation in India in recent times has suffered from several interrelated problems. Increased yields achieved during the green revolution period through input-intensive methods of high water and fertiliser use in well-endowed regions are showing signs of stagnation and concomitant environmental problems due to salinisation and waterlogging of fields. Water resources are also limited, and water for irrigation must contend with increasing industrial and urban needs. As a consequence of all these, rice farmers experienced a downturn in productivity growth (International Food Policy Research Institute 2009). Since rice plays a major role in fulfilling India's food security needs, the country may not be in a position to feed its population if the present growth rate and the current trend in the productivity of rice continue.

The average productivity of rice in India, at present, is 2.2 tons/ha, which is far below the global average of 2.7 tons/ha. The productivity of Indian rice is higher than that of Thailand and Pakistan but much lower than that of Japan, China, Vietnam and Indonesia. The downturn in productivity growth coupled with the increase in global food grain prices continues to threaten food security in many low-income countries. Between 2001 and 2007, global rice prices nearly doubled, primarily due to fall in supply caused by stagnation of yield (Gujja and Thiyagarajan 2009). Since there is not much scope of increasing the area of rice

P. Varma, *Rice Productivity and Food Security in India*,
DOI 10.1007/978-981-10-3692-7_1

cultivation, the additional production has to come from less land, less water and less of other inputs.

Therefore, any further increase in the production of rice depends heavily on the intensification of the available land under rice. However, the conventional system of production intensification based on fertiliser and other resource intensive methods had serious environmental implications such as depletion of water tables and erosion of soil fertility. The new intensification methods for rice cultivation are known as low external input sustainable agriculture, organic farming, ecological farming, intermittent irrigation, alternate wetting and drying, aerobic rice cultivation, etc. (Namara et al. 2003). The system of rice intensification (SRI) shares one or more of the aspects of these methods of rice cultivation.

Although there is considerable literature analysing the potential of SRI to enhance the yield, there is hardly any study analysing the factors influencing the depth and intensity of adoption of SRI at the microlevel for India. Intensity measures the number of acres devoted for SRI cultivation whereas depth measures the number of components of SRI being adopted. The few available studies on adoption of SRI and factors influencing the adoption of SRI in India have treated adoption as a binary dependant variable. Nonetheless, adoption is not simply a yes/no decision. The farmers can either adopt the entire package or some of the components. The factors influencing the intensity of SRI adoption can also vary from the factors influencing the depth of adoption. Since adoption is constrained by several factors, non-adopters can be expected to adopt this new technology only when the constraints are removed. In an imperfect market setting, even a farmer with positive demand for adoption may not able to adopt a new technology owing to several initial constraints such as lack of access to information and access to seed. (Shiferaw et al. 2008, 2015). There have hardly been any studies on SRI adoption by explicitly considering the effective information constraints in the adoption decision.

Adoption of technology, which is usually introduced as a package, goes through a long process in terms of full adoption. Farmers often choose only parts of a package or apply combination of practices (Smale et al. 1995). The set of explanatory variables that influence the decision to adopt one component can also influence the decision to adopt another component (Teklewold et al. 2013). Therefore, adoption decision is inherently multivariate. Most studies on adoption make use of the univariate modelling approach. This approach may, however, lead to inconsistent parameter estimation especially when there is interdependence and simultaneity in adoption between packages (Dorfman 1996; Teklewold et al. 2013).

Similarly, there are hardly any studies that rigorously analyse the impact of SRI on household income and yield in a joint framework of adoption and impact. Since the factors influencing the adoption can also have an impact on the outcome variables (yield and income), one need to address the endogeneity issues. The endogeneity issues can be addressed by analysing the factors influencing the adoption and the outcome variables in a joint framework. The multinomial endogenous treatment effects technique developed will address the issue of endogeneity. This methodology is discussed at length in Chap. 9.

Therefore, the present study contributes to the SRI literature by identifying the factors that influence the decision to adopt SRI and its impact on household income and yield.

1.2 Major Objectives of the Study

- To understand the need for sustainable rice farming practices in general and the SRI in particular.
- To make a detailed analysis of the factors affecting the adoption and the intensity of adoption of SRI among the selected rice-producing states of India.
- To study the factors influencing the adoption of different practices of SRI.
- To understand the regional variations in the nature of adoption of SRI.
- To analyse the impact of SRI adoption on rice yield and the household income.
- To understand the impact of government's National Food Security Mission (NFSM) on the promotion of SRI.
- To evaluate the usefulness of SRI in enhancing the rice productivity and food security in India.

1.3 Study Area, Data Collection and Methodology

Study Area

The states for analysis were carefully selected, taking into account the objectives of the study. The purpose of the study is to make a detailed analysis of factors (institutional, social, economic, agronomic, demographic, market, risk, etc.) that influence the adoption at different stages and the impact on household income and yield. As a first step, all the major rice-producing states are identified. It is found that there are around 13 major rice-producing states, and macrolevel studies have already been undertaken for these states (Palanisami et al. 2013). Thus, our analysis is focused at the microlevel. Given the importance of identifying states that have sufficient SRI cultivation, three states with the highest differences in yield and gross margin between traditional and SRI cultivation are selected based on the existing studies (Palanisami et al. 2013).

Then, there is the criterion of NFSM and non-NFSM districts. In order to further divide SRI districts into NFSM and non-NFSM districts, the states that we identify should have NFSM districts where SRI is promoted within the NFSM.

The SRI first took root in India primarily in the southern states of Tamil Nadu and Andhra Pradesh (Glover 2011). SRI initially diffused to Tamil Nadu, followed by Andhra Pradesh, and diffusion studies showed that the acceptance and spread of SRI was rapid in these two states (Johnson and Vijayaragavan 2011).

As far as the rice-producing south Indian states are concerned (Tamil Nadu, Karnataka, Andhra Pradesh and Kerala), Kerala has around six SRI districts but does not have any NFSM districts. The remaining three south Indian states have both SRI and SRI-NFSM districts. The yield differences and gross margin difference between traditional rice cultivation and SRI were one of the lowest in Tamil Nadu and highest in Karnataka (Palanisami et al. 2013). Therefore, out of the four south Indian states, the study selected Karnataka for the purpose of analysis.

Among the major rice-producing states from the western region (Gujarat, Rajasthan and Maharashtra), there were no SRI districts which were linked to NFSM. Moreover, even the number of SRI districts was relatively less as compared to other regions.

As for the states from the eastern region (Orissa, Chhattisgarh, Uttar Pradesh and West Bengal), the yield difference and the gross margin difference were the highest in Orissa. In the case of Chhattisgarh, the gross margin difference was very less (2%) even when the yield differences were relatively high (Palanisami et al. 2013). Due to the considerable gap in yield and gross margin between traditional and SRI cultivation in Orissa, the study makes use of the state for the purpose of analysis.

Madhya Pradesh, the only rice-producing state from the central region, had good number of SRI and SRI-NFSM districts. Also, the yield difference and the gross margin difference were high making it one of the highest among all the major rice-producing states.

The major rice-producing state from the north-eastern region (Assam) did not have many SRI and SRI-NFSM districts. The yield difference and gross margin difference were also at a low level.

Accordingly, three states are identified for the purpose of analysis based on the objectives of the study and the above criteria. They are Karnataka, Orissa and Madhya Pradesh. For the purpose of further identification of the districts, the study made use of three criteria, viz. agro-climatic zones, SRI districts and SRI-NFSM districts. Thus, two districts from each state that belong to the same agro-climatic zones were identified. Selection of the districts in each state belonging to same agro-climatic zones was done after ensuring that one is with SRI practice and the other is with SRI incorporated under NFSM.

For Karnataka, Hassan (SRI-NFSM) and Chikmagalur (SRI) districts are identified. Similarly for Orissa, Keonjhar (SRI-NFSM) and Mayurbhanj (SRI) districts are identified. For Madhya Pradesh, Sidhi (SRI) and Shahdol (SRI-NFSM) districts are selected. From each of these districts, the blocks/taluks were then selected. The selected blocks/taluks from each district were as follows: Alur, Hassan and Sakleshpur from Hassan; Chikmagalur from Chikmagalur; Sadar, Patna and Harichandapur from Keonjhar; Karanjia and Jashipur from Mayurbhanj; Sidhi and Sihawal from Sidhi; and Gohapru and Sohajpur from Shahdol.

Data Collection

This study is based on primary data collected through a comprehensive household survey in the above-mentioned six districts of three major rice-producing Indian states in the year 2015. The farmers were selected through a stratified random

Table 1.1 Overview of sample selection from the study area

State	District	SRI farmers	Non-SRI farmers	Total farmers
Karnataka	Chikmagalur	21	21	42
	Hassan	20	20	40
Madhya Pradesh	Shahdol	30	30	60
	Sidhi	32	32	64
Orissa	Keonjhar	49	49	98
	Mayurbhanj	41	41	82
Total	All	193	193	386

Source Author's survey data

sampling technique. The sample ensured the representation of almost all the available SRI farmers from the study region and equal number of non-adopters of SRI. The survey was conducted through questionnaire, carefully framed so as to draw out the details covering household characteristics, wealth and farm characteristics, institutional and access-related variables, risk and economic factors. The precise questions or variables included in the questionnaire were based on the insights received from various studies on agricultural technology adoption. The total number of farm households interviewed was 386, of which 193 households were SRI adopters (Table 1.1).

Methodology

The factors influencing the adoption of SRI (both in terms of intensity and depth) are analysed using a simultaneous equation model. The model will explicitly incorporate multiple binding constraints that farmers face while making adoption decisions. The methodology is discussed in detail in Chap. 7.

The analysis of factors influencing the adoption of multiple packages of SRI is done using multivariate probit as well as ordered probit models. A detailed discussion of this methodology is in Chap. 8.

Finally, the impact of adoption of SRI on household income and rice yield is analysed using a multinomial endogenous treatment effects model proposed by Deb and Trivedi (2006a, b). The detailed discussion of the methodology is given in Chap. 9.

1.4 Scope of the Study

Microstudies on technology diffusion among farmers are very important from policy perspective. The heterogeneity of agro-climatic and socio-economic factors can act as a hindrance in applying the learning from one place to another. Microstudies can provide useful background information about the farmers who are currently using a technology and those who are not. Without basic descriptive information on these two categories of farmers, it is difficult to know how to formulate policies aimed at improving agricultural productivity.

The study is intended to shed light into the factors influencing the adoption, intensity as well as the depth of adoption by explicitly considering information constraints in adoption. The study also analyses the factors influencing the adoption of multiple components of SRI. This is important especially since partial adoption of SRI is very common. From a policy perspective, it is very pertinent to understand region-specific constraints affecting the adoption of a technology and its components along with other socio-economic characteristics of farm households. Non-adoption of a technology by a farmer does not always indicate lack of interest in the technology; rather it could be due to several constraints such as lack of information, credit and seed. Therefore, even with a positive demand for adoption, a farmer may not be able to adopt a technology. Such insights are very crucial from a developing country perspective.

The study will also shed light into whether some of the government's policy interventions are effective and, if not, how to make them more effective. The selection of an NFSM district from each state is undertaken in order to meet this objective. Out of the two districts selected from each state, one district has SRI promotion through NFSM.

Another dimension of the study is to analyse the impact of SRI on yield and household income. Any positive impact of an adoption of a technology should lead to further sharpening of the policies in a more targeted manner.

1.5 Chapter Scheme

This study is divided into 10 chapters. This chapter as an introduction gives the background, objectives, data, methodology and the scope of the study along with chapter scheme. Chapter 2 provides an overview of rice economy. Chapter 3 discusses the role of sustainable rice production practices to improve rice productivity and food security with an emphasis on India. Chapter 4 discusses SRI in detail. Chapter 5 discusses the salient features of government of India's National Food Security Mission (NFSM) and its objectives especially in the context of rice production and SRI. Chapter 6 provides an overview of SRI adoption across different study regions along with a detailed discussion of the socio-economic profile of the sample households. Chapter 7 analyses the factors that influence the adoption of SRI under effective information constraints. Chapter 8 provides an in-depth understanding of factors that influence the adoption of multiple practices of SRI. Chapter 9 then analyses the impact of SRI adoption on household income and rice yield. Finally, concluding observations and policy implications are discussed in Chap. 10.

Chapter 2
An Overview of Rice Economy

2.1 Introduction

This chapter will provide an overview of Asian rice economy with a special emphasis on Indian rice economy. This chapter will highlight the recent performance of India by discussing the trends in rice production and productivity of rice in the states selected for the purpose of analysis. A detailed district-wise analysis of cropping pattern, irrigation status, etc., is also provided to get clarity on the study area. This chapter will also discuss the trends in rice trade and the export potential of Indian rice.

2.2 Asian Rice Scenario

Rice is the world's most important crop, and the food security in Asia has traditionally been defined as having stable prices for rice in the major urban markets of a country. Therefore, food security is essentially a reflection of rice security in this region. One-third of the economy in the Asian region is dependent on rice production, marketing and consumption. Rice is the staple food for more than 50% of the population in Asia, and for South Asia alone, the figure is around 70% (Bishwajit et al. 2013). Rice is the cheapest and most effective staple food crop available in this region that is likely to eradicate acute undernutrition. Additionally, more than half of the daily caloric intake in some countries comes from rice. Asia is the largest producer of rice contributing around 91% of total world rice production (see Fig. 2.1). The remaining rice production is divided between Africa and Latin America. However, rice is gaining importance in these regions with a 40% increase in annual per capita consumption of rice in Africa and 46% increase in annual per

P. Varma, *Rice Productivity and Food Security in India*,
DOI 10.1007/978-981-10-3692-7_2

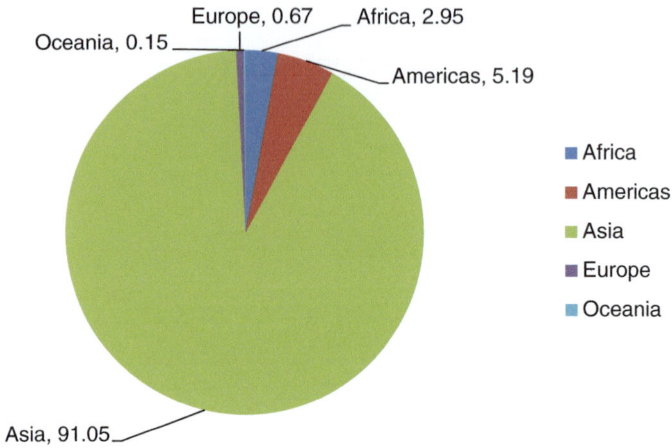

Fig. 2.1 Rice cultivation in the world. *Source* FAOSTAT (average of 1980–2014)

capita consumption of rice in Latin America (Yamano et al. 2016). The rice consumption in Africa increased from 16.7 kg in 1990 to 23.3 kg in 2011. Similarly, the rice consumption in Latin America increased from 7.1 kg in 1990 to 10.4 kg in 2011. This shift in consumption pattern also indicates that the global rice production will need to increase by 116 million tons (26%) by 2035, from the 2010 production level, to meet the rising demand (Yamano et al. 2016).

Several studies have revealed that there is a great potential to further increase the rice production mainly in South Asian countries (Bishwajit et al. 2013). The introduction of modern breeding strategies, which has proved to be much more efficient than conventional techniques and environmental stress-tolerant varieties, offers considerable scope for the increase in rice production in the South Asian region. In South Asia, Bangladesh has had the highest rice yield since 1992. India's rice yield remains lower than that of Bangladesh and Sri Lanka and higher than that of Pakistan and Nepal. Despite having high production rate, Bangladesh imports rice as it is the major consumer of rice in the South Asian region. India is the largest exporter and the fourth largest rice-consuming nation in the South Asian region (Bishwajit et al. 2013).

Among the South Asian countries, rice occupies special position in the Indian economy. Rice is one of the most important food crops of India. India is the second largest producer of rice after China (see Fig. 2.2). It plays a major role in India's diet and is central to food security. Rice is the staple food crop for more than 70% of the Indian people. Ninety per cent of the rice produced is consumed within the country. Major share of rice is cultivated during Kharif season. A small share of rice is grown in Rabi/summer season with assured irrigation.

Rice plays a special role in Indian tradition. For Indians, rice is more than mere livelihood; it has shaped the history, culture, art and lifestyle of its population in many ways.

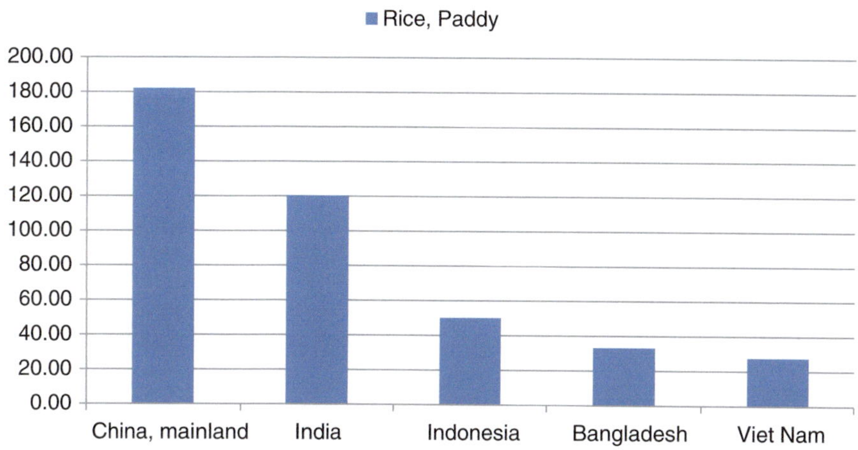

Fig. 2.2 Major rice-producing countries (in million tonnes). *Source* FAOSTAT (average of 1980–2014)

2.3 Rice Scenario in India

Rice production in India increased from around 80 million tonnes in 1980 to around 157 million tonnes in the year 2014 (see Fig. 2.3). Productivity also improved from around 20 hg per ha to around 36 hg per ha during 1980–2014 (see Fig. 2.4). Nonetheless, the area under cultivation has not increased much; it was around 40 million ha in 1980 and 43 million ha in the year 2014 (see Fig. 2.5). Although the area under cultivation has increased marginally, the trends in area show that the future production of rice may face some challenges.

As per the FAO estimates the annual growth rates of area under cultivation, production and yield were −0.17, 2.04 and 2.2, respectively (see Fig. 2.6). The negative growth rates for area under cultivation indicate that any further increase in production of rice is possible only through methods that enhance rice yield substantially. Although India's rice yield was showing a marginal improvement over the years, the current rice yield is much lower than other countries (see Fig. 2.7). As mentioned earlier, India's rice yield remains lower than other South Asian counties such as Bangladesh and Sri Lanka as well (see Fig. 2.8).

On the trade front, from being a rice importer in the initial years of the 1980s, India transformed itself with a production of 157 million tonnes and exports around 11 million tonnes (see Fig. 2.9). India is the largest exporter of rice in the world market since 2011. From the 1960s to the 1980s, the USA was the largest exporter of rice with Thailand occasionally shipping more than the USA (Lakkakula et al. 2015). Gradually, Thailand became the largest exporter in 1981 and in the mid-1990s; Vietnam began to register large rice trade surpluses as a result of market and trade reforms (Lakkakula et al. 2015). India emerged as one of the major players in the world rice market in the recent years (Varma and Issar 2016). India's

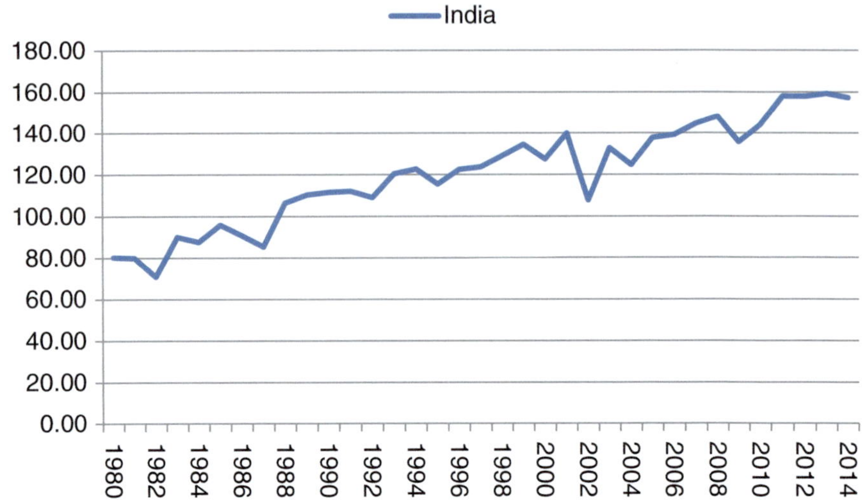

Fig. 2.3 Trends in rice production in India, in million tonnes. *Source* FAOSTAT (average of 1980–2014)

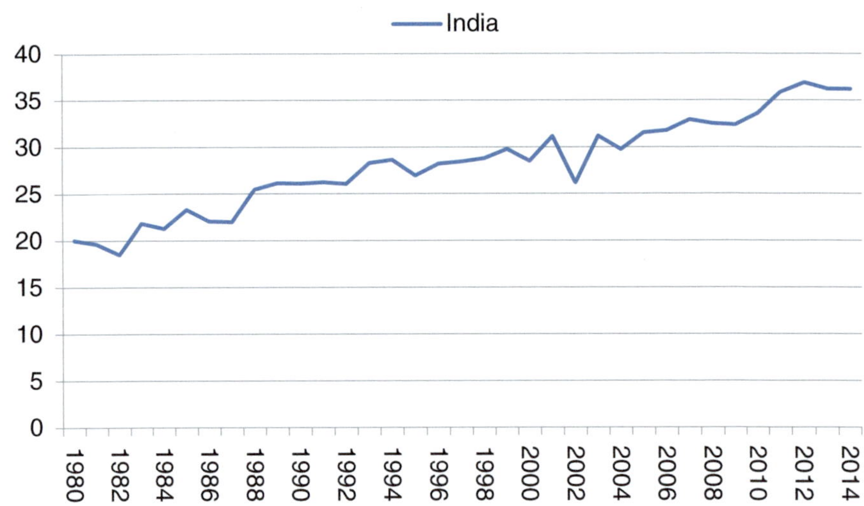

Fig. 2.4 The trends in yield of rice in India (in hg/ha). *Source* FAOSTAT (average of 1980–2014)

share in total world rice exports was around 26% in 2011–12 but marginally declined to 22% in 2013–14 (see Table 2.1). In 2011–12, Thailand's share was only 18%, but in 2013–14, the share increased to 22%. So in 2013–14, both India and Thailand had the same share of rice exports in the world market showing the severe competition between the two countries. Pakistan and China are the two other major rice players in the world market.

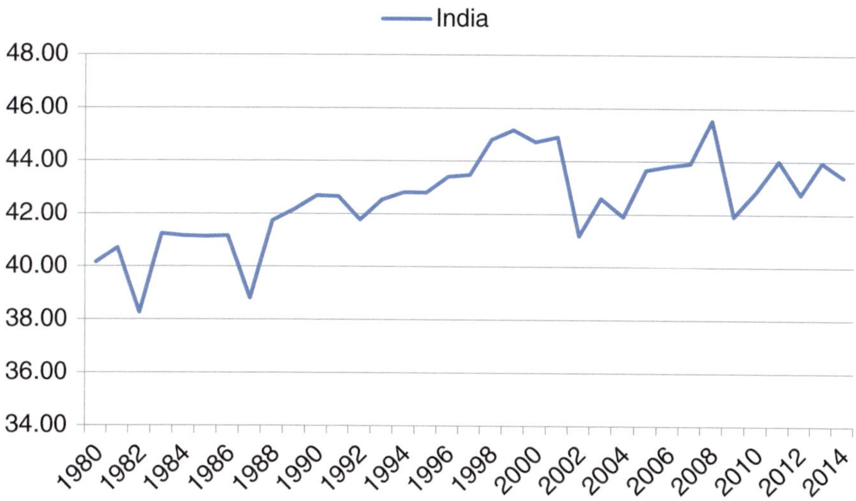

Fig. 2.5 Area under rice in India (in million ha). *Source* FAOSTAT (average of 1980 to 2014)

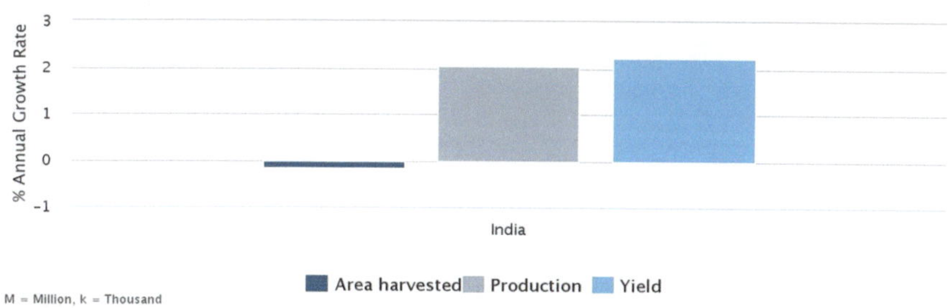

Fig. 2.6 Annual growth rates for area, production and yield of rice in India. *Source* FAOSTAT

India exports two varieties (basmati and non-basmati) of rice to the world market. The major importing countries for these two varieties of rice are also different. Basmati rice is exported to Iran, Saudi Arabia, Iraq, etc. whereas non-basmati rice is exported to Benin, Senegal and Bangladesh (see Tables 2.2 and 2.3). Among the two varieties of rice, the performance of non-basmati rice is remarkable as it almost doubled between 2011–12 and 2013–14 from 3.9 million tonnes to 7.13 million tonnes (Varma and Issar 2016). The global norms of the Food and Agriculture Organisation (FAO) are also beneficial to the Indian rice exporters. According to the Codex standard, the maximum acceptable level of arsenic in rice is less than 0.2 mg/kg in rice. Since the arsenic level in countries like the USA is more than 0.3 mg/kg of arsenic, India has an advantage in the world market in terms of meeting the Codex standard (Jha 2014).

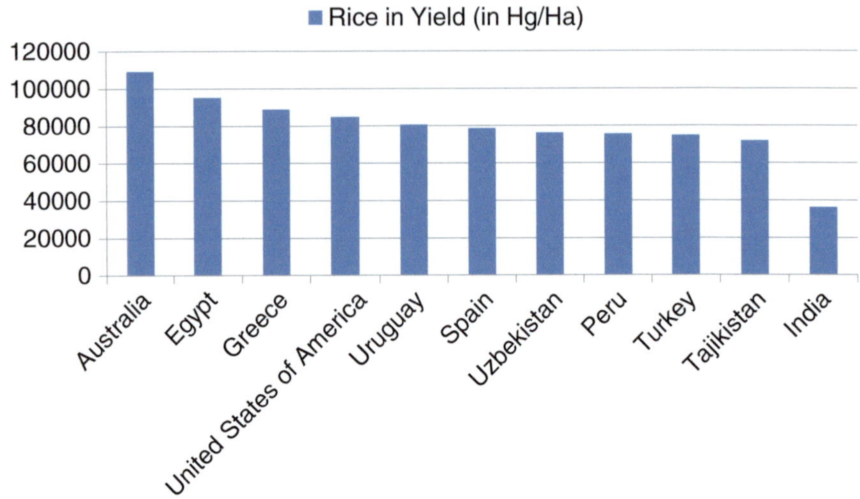

Fig. 2.7 India's position in yield of rice (paddy) in the world in 2014 (in hg/ha). *Source* FAOSTAT

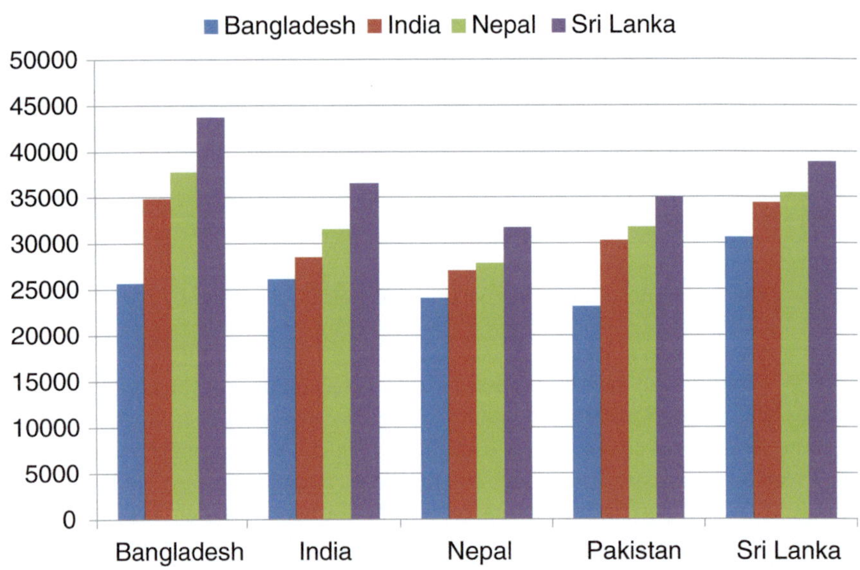

Fig. 2.8 India's rice yield in comparison with selected South Asian countries (in hg/ha). *Source* FAOSTAT

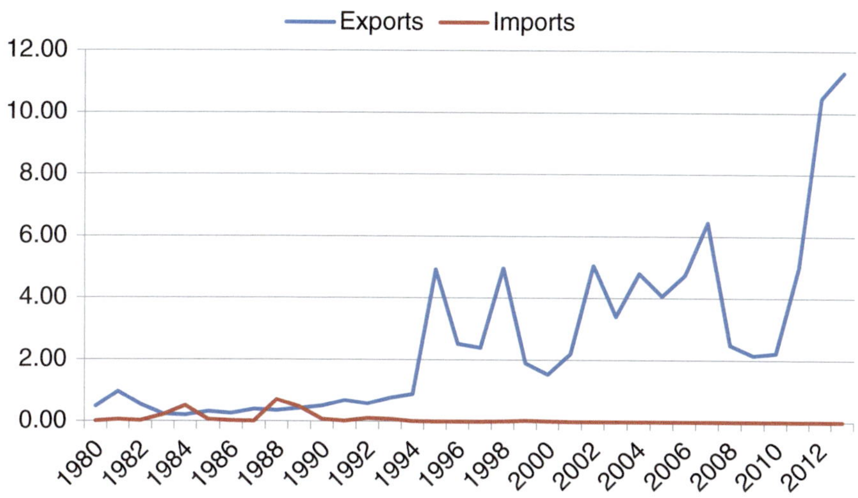

Fig. 2.9 Rice exports and imports in India (in million tonnes). *Source* FAOSTAT

Table 2.1 Rice trade and its potential (% share in total world exports)

Countries	2008–09	2009–10	2010–11	2011–12	2012–13	2013–14
Argentina	2.02	1.48	2.02	1.55	1.55	1.62
Australia	0.06	0.17	0.86	1.15	1.30	1.35
Brazil	2.01	1.36	3.58	2.82	2.20	2.20
Burma	3.58	1.41	2.15	1.76	1.94	1.94
Cambodia	2.79	2.38	2.37	2.04	2.53	2.59
China	2.66	1.96	1.34	0.68	0.78	0.52
Egypt	1.96	1.81	0.88	1.53	2.20	2.20
Guyana	0.82	0.94	0.71	0.73	0.78	0.84
India	7.31	7.06	12.79	26.18	23.31	22.00
Pakistan	10.84	12.67	9.42	8.68	8.29	8.28
Paraguay	0.42	0.43	0.57	0.67	0.65	0.65
Russia	0.31	0.49	0.39	0.72	0.54	0.60
Thailand	29.15	28.66	29.37	17.74	20.72	22.00
Uruguay	3.15	2.56	2.32	2.70	2.33	2.33
Vietnam	20.24	21.33	19.31	19.71	19.17	19.93
Others	2.42	3.04	2.95	2.84	2.89	2.68
USA	10.26	12.25	8.96	8.50	8.81	8.28

Source FAOSTAT

Table 2.2 India's basmati rice exports to major destination markets (in %), 2008–09 to 2014–15

Country/year	2008–09	2009–10	2010–11	2011–12	2012–13	2013–14	2014–15[a]
Canada	1.15	0.46	0.72	0.90	0.80	0.77	0.53
Iran	10.98	18.10	19.26	19.35	31.31	38.34	23.18
Kuwait	7.17	6.92	7.31	6.29	4.72	4.67	4.42
Netherland	0.72	0.22	0.99	1.17	1.74	1.16	1.59
Oman	1.61	0.29	0.25	0.58	1.16	1.15	1.27
Saudi Arabia	33.69	31.79	26.27	22.69	19.70	21.99	29.90
United Arab Emirates	29.31	30.58	27.99	22.93	6.77	3.97	4.16
UK	4.02	1.83	3.28	4.46	5.56	3.16	3.78
USA	2.55	1.29	1.93	2.89	2.64	2.75	3.05
Yemen Republic	2.02	3.07	2.85	2.90	4.98	3.91	4.54
Total	93.22	94.54	90.84	84.15	79.38	81.87	76.41

Source Own calculation based on APEDA data
Notes [a]Data is till June 2014

Table 2.3 India's non-basmati rice exports to major destination markets (in %), 2011–12 to 2014–15

Country/year	2011–12	2012–13	2013–14
Benin	5.35	8.62	16.35
Bangladesh	3.62	0.47	9.14
Senegal	8.35	12.78	9.14
South Africa	4.81	6.44	5.53
Liberia	0.78	2.49	4.53
Nepal	3.52	5.93	5.67
United Arab Emirates	5.19	3.92	3.15
Coast of Ivory	7.06	8.92	3.67
Madagascar	0.87	1.53	3.52
Guinea	0.38	4.21	3.11
Saudi Arabia	2.57	2.14	1.95
Cameroon	1.39	3.11	3.06
Nigeria	20.36	12.16	2.70
Singapore	1.47	2.01	1.87
Indonesia	5.76	2.03	1.69
Total (%)	71.48	76.77	75.07

Source Own calculation based on APEDA data

2.4 Area, Production and Yield of Rice—Selected States

Orissa

Orissa is a major rice-producing state of India. Rice covers around 69% of the total cultivated area and about 63% of the total area under food grains (Das 2012). Rice is the staple food of almost the entire population of the state, and therefore, the state economy is directly linked with improvements in production and productivity of rice in the state (Das 2012). Although Orissa was doing well in terms of rice production in the 1950s by supplying major portion of the rice grain in the central pool of food stocks, its situation was strongly reversed in the post-high-yielding variety period. The state's rice area has stagnated and its share in the country's production of rice declined from 11% in the pre-HYV period to 7.9% in the post pre-HYV period (Das 2012). In fact, the introduction of high-yielding varieties did not have any perceptible impact on rice production and productivity for over two decades: per hectare yield fluctuated between 800 and 977 kg (Das 2012). The comparison of average yield in three different time periods shows that the yield increased substantially from 1372 kg per hectare in the first period (2001–02 to 2005–06) to 1591 kg per hectare in the second period (2006–07 to 2010–11), while the yield remained almost stagnant in the latter two periods (2006–07 to 2010–11 and 2008–09 to 2012–13). The all India average rice yield increased from 2001 kg per hectare in the first period to 2175 in the second period and to 2279 in the third period. Although the average rice yield marginally increased for India, the rate of increase does not appear to be satisfactory. The average rice yield in Orissa was much lower than the average yield of India in all the three periods analysed (see Fig. 2.10).

In Orissa, rice is cultivated in an area of 4.45 million hectares (Das 2012). The rice cultivation can be classified into six different ecosystems, namely irrigated

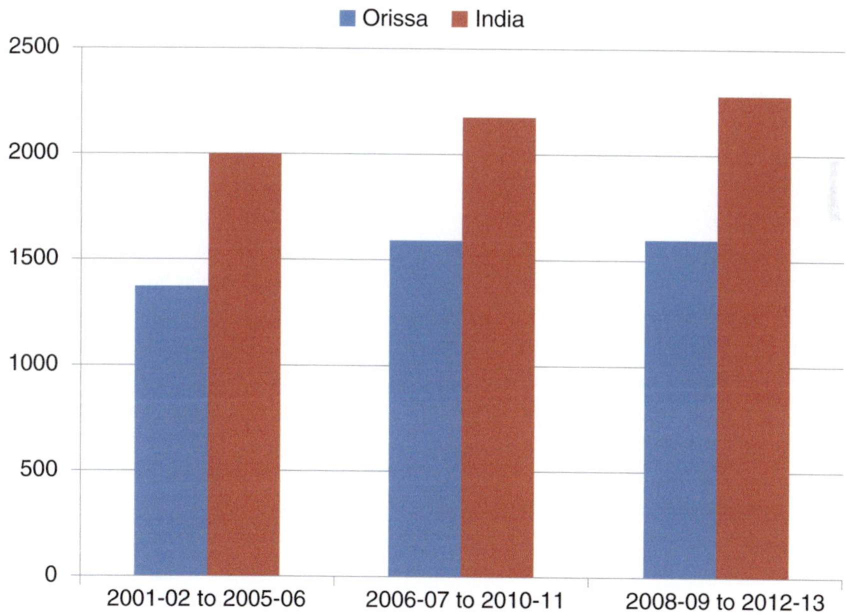

Fig. 2.10 Average rice yield in Orissa and India for different periods (in kg/ha). *Source* Directorate of Economics and Statistics, Ministry of Agriculture

Kharif (27.4%), rain-fed upland (19.1%), medium land (12.4%), shallow lowland (22.5%), semi-deep (7.9%), deep (3.4%) and irrigated Rabi (7.4%) (Das 2012).

Rice is grown during winter, summer and autumn seasons. Winter rice or Kharif rice is the dominant crop accounting for 77% of total rice produced in the state (Pandey et al. 2012). During Kharif season planting starts in June–August and harvest occurs in December–January. Autumn rice which is also known as pre-Kharif rice is cultivated during May–June to September–October. This accounts for around 11% of the total rice produced in the state and is usually cultivated in the rain-fed upland areas (Pandey et al. 2012).

The production of rice in the state increased from 3853 million tonnes in 1981–82 to 7613 million tonnes in 2013–14. Although there was an increase in production throughout the period, the data shows that the production fluctuated highly (see Table 2.4). During the same period, the all India production of rice was steadily increasing (see Table 2.4). All India production of rice increased from 53,250 million tonnes in 1981–82 to 106,540 million tonnes in 2013–14. Orissa's share in India's rice production remained stagnant during the period of 1980–81 to 2013–14. The area under rice also remained stagnant during the same period (see Table 2.5). The same was the case with the share of Orissa's rice area in total rice area; in fact, there was a marginal decline in the share of Orissa's rice area in total area.

The production and productivity growth rate trends of rice in Orissa indicated that the introduction of high-yielding varieties did not have any perceptible impact over two decades since their introduction during the mid-1960s and productivity remained below 1 tons/ha (Das 2012). During 2008–09, 77% of the total rice area (3.43 million hectares) was under high-yielding varieties whereas 75% (3.09 million hectares) was covered by high-yielding varieties during the Kharif season and the rest was under local and improved varieties (Das 2012) (Table 2.6).

Madhya Pradesh

The area under rice in Madhya Pradesh declined drastically from 4850 thousand hectares to 1930 thousand hectares in the period of 1981–82 to 2013–14. It was only Madhya Pradesh that witnessed a sharp decline in the area under rice cultivation; the area under rice cultivation in the other two states was almost stagnant. The increase in area under rice cultivation was very negligible at the all India level as well (see Table 2.5). The production of rice also declined from 3830 million tonnes in 1981–82 to 2845 million tonnes in 2013–14. In contrast, the production showed an increase both at the national level and in the other two states (see Table 2.4). But what is really striking here is that although the production of rice in Orissa and Karnataka showed an increase in a highly fluctuating manner, the national production showed a steady and consistent increase.

The increase in the rice yield was the highest in Madhya Pradesh. The yield increased from 790 kg per hectares in 1981–82 to 2666 kg per hectares in 2013–14. Interestingly, the increase in rice yield was mainly due to the reduction in area rather than an increase in production. The rice yield in Madhya Pradesh was much lower than the national average as well (see Fig. 2.11). Although rice was

Table 2.4 State-wise production of rice in million tonnes

States	1981–82	1985–86	1990–91	1995–96	2000–01	2005–06	2010–11	2011–12	2012–13	2013–14
Karnataka	2364	1943	2415	3024	3847	5744	4188	3955	3364	3573
Madhya Pradesh	3830	5418	5738	5839	982	1656	1772	2227	2775	2845
Orissa	3853	5226	5275	6226	4614	6859	6828	5807	7296	7613
Total of 3 states	10,047	12,587	13,428	15,089	9443	14,259	12,788	11,989	13,435	14,031
All India	53,250	63,830	74,290	76,980	84,980	91,790	95,980	105,300	105,240	106,540

Source Directorate of Economics and Statistics, Ministry of Agriculture

Table 2.5 State-wise area under rice in 1000 ha

States	1981–82	1985–86	1990–91	1995–96	2000–01	2005–06	2010–11	2011–12	2012–13	2013–14
Karnataka	1168	1096	1173	1265	1483	1485	1540	1416	1278	1340
Madhya Pradesh	4850	5032	5118	5344	1708	1658	1603	1662	1883	1930
Orissa	4159	4402	4404	4529	4434	4479	4226	4005	4023	4180
Total of 3 states	10,177	10,530	10,695	11,139	7625	7622	7369	7083	7183	7450
All India	40,710	41,140	42,690	42,840	44,710	43,660	42,860	44,010	42,750	43,950

Source Directorate of Economics and Statistics, Ministry of Agriculture

Table 2.6 State-wise yield of rice in kg per ha

States	1981–82	1985–86	1990–91	1995–96	2000–01	2005–06	2010–11	2011–12	2012–13	2013–14
Karnataka	2025	1772	2059	2390	2593	3868	2719	2793	2632	2666
Madhya Pradesh	790	1077	1121	1093	575	999	1106	1340	1474	1474
Orissa	926	1187	1198	1375	1041	1531	1616	1450	1814	1821
All India	1308	1552	1740	1797	1901	2102	2239	2393	2462	2424

Source Directorate of Economics and Statistics, Ministry of Agriculture

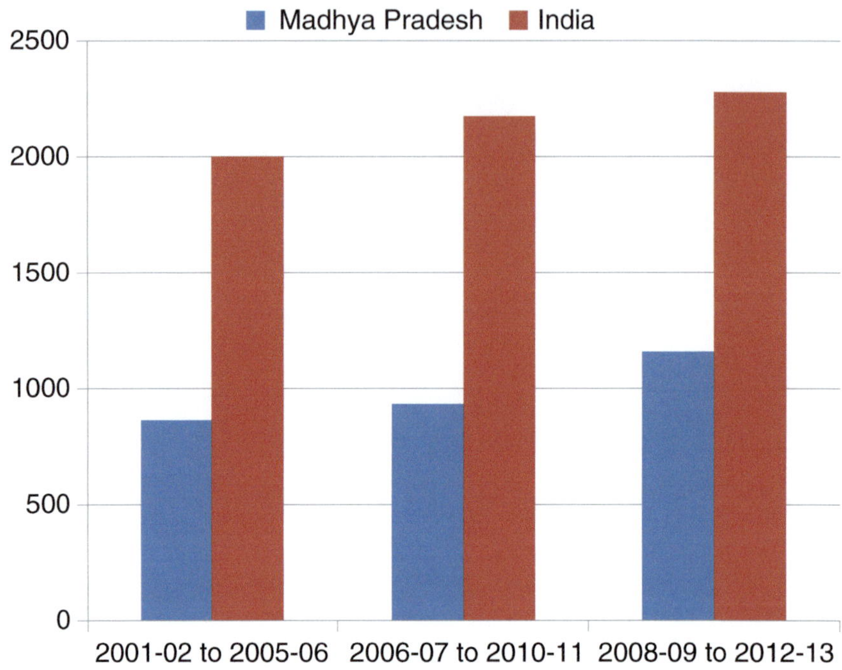

Fig. 2.11 Average rice yield in Madhya Pradesh and India for different periods (in kg/ha). *Source* Directorate of Economics and Statistics, Ministry of Agriculture

incorporated under the government of India's National Food Security Mission (NFSM) to improve the rice production, the productivity of rice was found to be higher in non-NFSM districts of Madhya Pradesh as compared to NFSM districts (Agro Economic Research Centre for Madhya Pradesh and Chhattisgarh Report 2013).

Karnataka

Karnataka is a major rice-producing state of India and rice is grown under a variety of soils and wide range of rainfall and temperature. Rice production in Karnataka depends heavily on monsoon and only 44% of the total rice acreage is under irrigation (Rajanna 2010). The unique feature of rice cultivation in the state is that either sowing or transplanting is seen in all seasons of the year (Rajanna 2010). The duration of the rice varieties cultivated in the state varies from 100 to 180 days depending on season and agro-climatic location, and the rice-growing ecosystems of the state can be broadly divided into six categories (Rajanna 2010). They are coastal area, hilly area, transitional area, tank-fed area, irrigated maidan area (south) and irrigated maidan area (north). In our analysis, we have selected Hassan and Chikmagalur districts from Karnataka. Hassan belongs to irrigated maidan (south) area whereas Chikmagalur is a hilly area.

Kharif (June–July) and summer (January–February) are the two main rice seasons of the state. There are six rice-growing ecosystems in Karnataka. Out of this, Kharif sowing is more common. In summer, rice is cultivated mainly in the irrigated maidan areas of north and south. In coastal area, crop is sown in September–October and harvested in January–February and then another crop sowing season is in December–January and the harvesting is in March–April. In the tank-fed areas, the crop cultivation is in the months of August–September depending upon the arrival of monsoon. In each district around 60–80% of the total area is devoted for Kharif while the remaining is devoted for late Kharif and summer seasons (Rajanna 2010). The state also has plenty of traditional rice varieties cultivated. The most popular traditional varieties cultivated are Rajaboga, Rajamudi, Kayame, Ratnachudi and Jerasanna.

Among the three states, rice productivity was the highest in Karnataka—higher even than the national average (see Fig. 2.12). But the average yield showed a marginal decline in the later period. The average yield was 2712 kg per hectares during 2001–02 to 2005–06 and declined to 2626 kg per hectare during 2008–09 to 2012–03 (see Fig. 2.11). During this period, the all India average yield showed a marginal improvement.

The production of rice increased from 2364 million tonnes in 1981–82 to 5744 million tonnes in 2005–06 but thereafter witnessed a decline to 3573 million

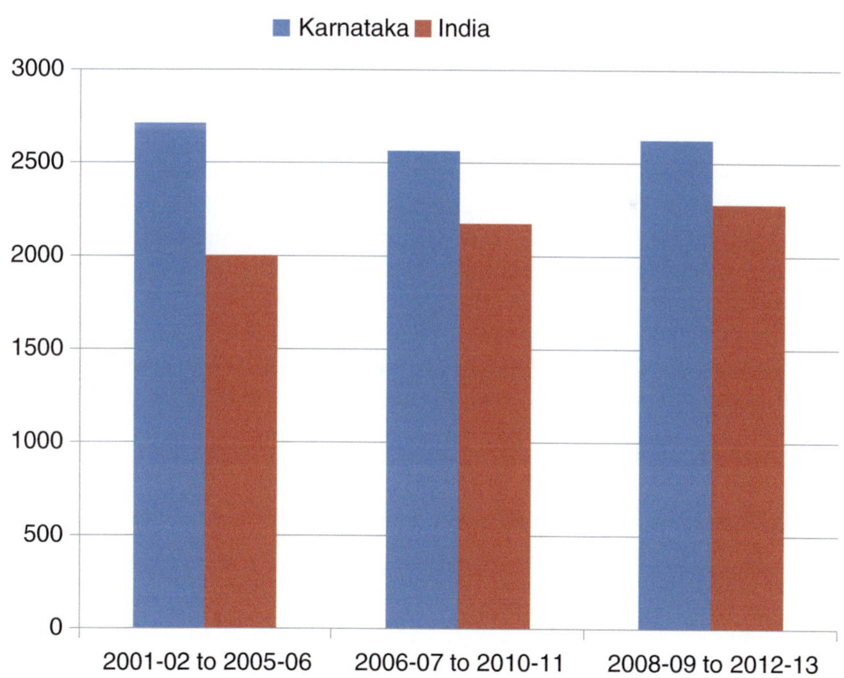

Fig. 2.12 Average rice yield in Karnataka and India for different periods (in kg/ha). *Source* Directorate of Economics and Statistics, Ministry of Agriculture

tonnes in 2013–14 (see Table 2.3). The area under rice cultivation increased marginally from 1168 thousand hectares in 1981–82 to 1340 thousand hectares in 2013–14 (see Table 2.4).

2.5 District-Wise Profiles

Karnataka

Hassan belongs to southern dry agro-climatic zone and has red sandy loams in major areas and pockets of black soil in remaining areas. On the other hand, Chikmagalur belongs to the hilly zone with red clay loamy soils in major areas. Rainfall in Hassan district ranges from 671 to 889 mm and from 904 to 3695 mm in Chikmagalur district of Karnataka. Important Kharif crops in Hassan district include paddy, ragi, jowar, maize, tur, H gram, cowpea, avare, groundnut, sesamum, sunflower, castor, niger, cotton, tobacco and sugar cane. Important Rabi crops in Hassan district are ragi, maize, H gram and cowpea. In Chikmagalur, paddy, cotton and sugar cane are important Kharif crops whereas paddy, black gram and green gram are important Rabi crops in the district.

Hassan accounts for 2.40 area in lakh hectares under Kharif crops and 0.46 lakh hectares of area under Rabi crops. On the other hand, Chikmagalur accounts for 1.53 lakh hectares of area under Kharif crops and 0.22 lakh hectares of area under Rabi crops.

The data on number and area of operational land holders in Karnataka shows that the selected districts—Chikmagalur and Hassan—had high number of marginal farmers (30,056 in Chikmagalur and 52,968 in Hassan) (see Table 2.7). As a result, the percentage share of small farmers in total farmers was also the highest in Hassan (72.4%) (see Figs. 2.13 and 2.14). Although the number of small farmers was higher in Hassan, in percentage terms Chikmagalur had highest share of small farmers (22.4%); in Chikmagalur, it was 20.7%. The percentage shares of semi-medium, medium and large farmers were also the highest in Chikmagalur.

The area irrigated by canals, tanks, wells, tube wells and lift irrigation was the highest in Hassan as compared to Chikmagalur. The total net irrigated area under cultivation was the highest in Hassan as compared to Chikmagalur (see Table 2.8).

Table 2.7 Number and area of operational land holders in study districts of Karnataka (holdings in no. of farmers, area in hectares)

District	Marginal farmers		Small farmers		Semi-med farmers		Medium farmers		Large farmers	
	No.	Area	No.	Area	No.	Area	No.	Area	No.	Area
Chikmagalur	30,056	13,346	10,838	14,575	4655	12,135	2079	12,035	692	16,401
Hassan	52,968	26,346	15,136	20,894	4249	10,969	733	3876	41	726

Source Agricultural census 2010–11, Directorate of Economics and Statistics, Bangalore

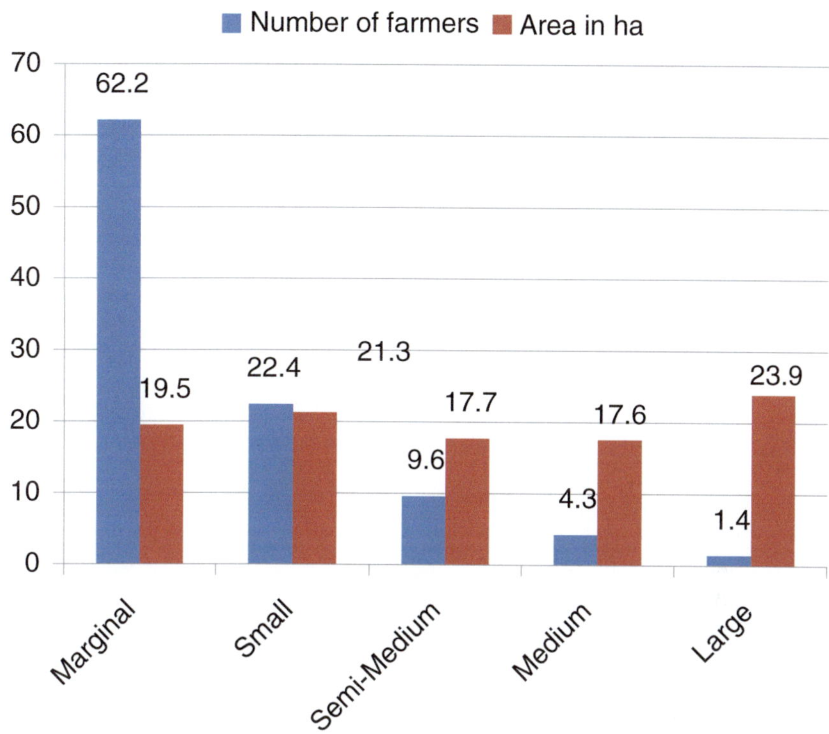

Fig. 2.13 Percentage of number and area of operational land holders in Chikmagalur. *Source* Agricultural census 2010–11, Directorate of Economics and Statistics, Bangalore

In Hassan district, the total area sown during Kharif was 2.38 lakh hectares out of which 1.97 lakh hectares was rain-fed, and irrigated area accounted for 0.41 lakh hectares. However, area affected by dry spell accounted for 0.61 lakh hectares. Total area sown and area affected by dry spell in Chikmagalur district during Kharif was relatively less compared to the respective areas in Hassan (see Table 2.9).

Madhya Pradesh

Madhya Pradesh is endowed with 11 agro-climatic zones, 5 crop zones and 7 soil types. The Sidhi district has red and yellow soil type whereas the Shahdol district has red and yellow medium black soil type. As far as the area under cultivation is concerned food grain cultivation is more prominent with 60% of the total cropped area. The remaining area is devoted for the cultivation of oilseeds, fibre crops and other cash crops. The main crops cultivated in the state are rice, wheat, maize and jowar and cereals, gram and lentil in pulses and soya bean and mustard in oilseeds. Around 63% of the total cropped area is devoted for Kharif crops and 37% is devoted for Rabi crops. Out of the total cropped area, 38% is under cereals, 22% is under pulses, 31% is under oilseeds and the remaining is under vegetables, fruits and other crops.

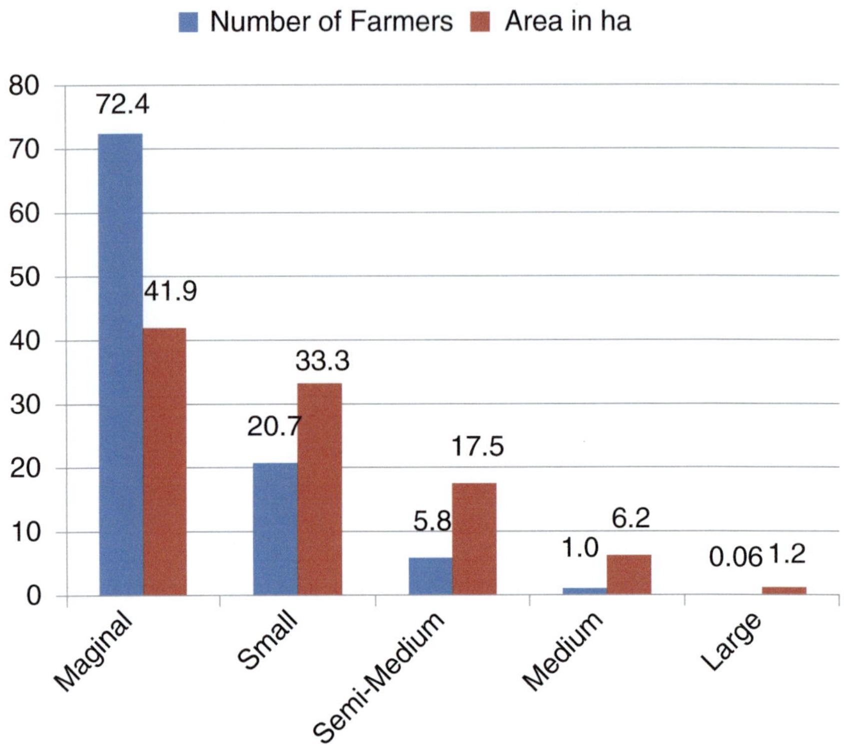

Fig. 2.14 Percentage of number and area of operational land holders in Hassan. *Source* Agricultural census 2010–11, Directorate of Economics and Statistics, Bangalore

Table 2.8 Source-wise net irrigated area in study districts of Karnataka, 2010–11 (area in hectare)

District	Canals	Tanks	Wells	Tube/borewells	Lift irrigation	Other sources	Total
Chikmagalur	4666	10,822	637	10,977	192	11,945	39,239
Hassan	45,398	30,140	1719	47,294	325	1657	126,533

Source Annual season and crop statistics report 2010–11 of DE&S, Bangalore

Table 2.9 Targeted area and area sown in study districts of Karnataka, Kharif 2012–13 (area in lakh hectare)

District	Targeted area	Area sown			% coverage	Unsown area	Area affected by dry spell
		Irrigated	Rain-fed	total			
Hassan	2.55	0.41	1.97	2.38	93	0.17	0.61
Chikmagalur	1.62	0.1	1.28	1.37	85	0.24	0.28
State	74.7	17.25	45.63	62.88	84	11.81	16.21

Source Karnataka state statement-II http://raitamitra.kar.nic.in/imp_agri_stat.html)

There has been a considerable change in the cropping pattern of Madhya Pradesh in the past two to three decades. Area under cereals has declined over time. While area under wheat and maize has gone up, area under jowar and other cereals has declined. In pulses, area under gram has gone up by nearly 65%, and Madhya Pradesh today is the largest producer of gram in the country. Area under soya bean observed an enormous growth between 1978 and 2006–07, a rise by almost 23 times.

The total production of crops in both Kharif and Rabi season was slightly higher in Sidhi as compared to Shahdol (see Table 2.9). Although Kharif yield was much higher in Sidhi district, the Rabi yield was slightly higher in Shahdol district. Paddy and wheat were the main crops cultivated in Shahdol, whereas paddy, wheat, maize, tur, sesamum, barley, gram, etc., were the main crops cultivated in Sidhi district (see Table 2.10).

Irrigation Status

Wells and tube wells are the major source of irrigation covering almost 66% of net irrigated area in Madhya Pradesh. The net area sown in Shahdol district of Madhya Pradesh is 179,700 ha, out of which the gross irrigated area is 21,000 ha. However, the rain-fed area accounts for 185,800 ha. In Sidhi district, out of 353,800 ha of net area sown, only 66,800 ha is gross irrigated and 413,500 ha is rain-fed area (see Table 2.11).

In Madhya Pradesh, the number of small and marginal farmers is about 62% of total land holders, but their share in area is only 22%.

Orissa

Orissa has a geographical area of 155,707 km^2 the state has 10 agro-climatic zones based on soil types, topography, rainfall and cropping pattern. Like the other two

Table 2.10 Crop production, productivity and cropping pattern in study districts of Madhya Pradesh

District	Name of crops	Kharif			Rabi		
		Area ('000 ha)	Average yield (kg per ha)	Production ('000 tonnes)	Area ('000 ha)	Average yield (kg) per ha	Production ('000 tonnes)
Shahdol	Paddy, kodokutki, wheat	165.9	80	114.6	34.5	646	22.3
Sidhi	Paddy, maize, kodokutki, tur, sesamum, wheat, barley, gram, linseed	297.3	670	200.1	178.6	629	112.3

Source Madhya Pradesh, Department of Land Resources

Table 2.11 Source-wise irrigation status in study districts of Madhya Pradesh (in hectare)

District	Canals	Tanks	Tube wells	Wells	Other sources	Gross irrigated area
Shahdol	4000	900	2600	1600	11,900	21,000
Sidhi	12,800	400	26,500	11,000	16,100	66,800
Madhya Pradesh	990,500	135,200	2,294,500	1,534,800	825,100	63,708

Source Madhya Pradesh, Department of Land Resources

Table 2.12 Number, area and average size of operational holdings in study district of Orissa, 2010–11

District	No. of holdings	Operated area (in ha)	Average size of holding
Keonjhar	256,477	253,167	0.99
Mayurbhanj	389,981	357,627	0.92
State total	4,667,466	4,852,014	1.04

Source Directorate of Economics and Statistics, Odisha

states, majority of farmers are small and marginal with low levels of literacy. The total number of operational holdings is 43.56 lakh. The operational area as per the agricultural census 2005–06 is 50.19 lakh hectares. The state has tropical climate with high temperature and humidity. The rainfall availability is medium and winter is very mild. The gross cropped area in the year 2011–12 was 88.01 lakh hectare, and the cropping intensity was 166%. The major crop cultivated during the Kharif season is rice whereas the major crop cultivated during the Rabi season is pulses and oilseeds (Directorate of Agriculture & Food Production, Odisha 2011–12).

There are couple of government schemes that are being implemented in the state. They include NFSM, System of Rice Intensification (SRI) and Rashtriya Krishi Vikas Yojana (RKVY). The main objective of all these programmes is to increase the crop production and the productivity of various crops.

Keonjhar and Mayurbhanj districts of Orissa lie in the North Central Plateau agro-climatic zone. The climate in Keonjhar district is dry subhumid, and in Mayurbhanj district, it is moist subhumid.

As per the Agricultural Census 2010–11, there were a total of 46.67 lakh operational holdings in the state. The operational holdings in the study districts—Keonjhar and Mayurbhanj—were around 2.56 lakh and 3.89 lakh, respectively (see Table 2.12). The operated area in hectare in the state was around 48 lakh hectare, and for Keonjhar and Mayurbhanj, the operated areas were 2.53 lakh hectares and 3.57 lakh hectare, respectively (see Table 2.12). The average size of holding in marginal, small, semi-medium, medium and large categories in 2010–11 was 0.57, 1.63, 2.95, 5.99 and 23.72 ha, respectively (Directorate of Economics and Statistics, Orissa, 2010–11).

In Madhya Pradesh, marginal farmers owned around 19.21 lakh hectares, small famers owned around 14.97 lakh hectares, semi-medium farmers owned 9.18 lakh hectares, medium farmers owned 3.81 lakh hectares and large farmers owned 1.32 lakh hectares (see Table 2.13). The total area under all the categories of farmer groups was relatively higher in Mayurbhanj as compared to Keonjhar.

The agro-climatic characterisation of the two districts showed that annual rainfall was around 1484 mm in Keonjhar and 1553 mm in Mayurbhanj (see Table 2.13). The percentage share of net irrigated area in total area was 25% in Keonjhar whereas it was 29% in Mayurbhanj (see Table 2.14).

The gross cropping intensity in Madhya Pradesh was 166% whereas in Keonjhar and Mayurbhanj the same was 152 and 143%, respectively. The gross cropped area and the net sown area was relatively higher in Mayurbhanj as compared to Keonjhar (see Table 2.15).

The irrigation status of the state showed that the net irrigated area and gross irrigated area were marginally higher in Mayurbhanj as compared to Keonjhar. The total cropped area was also higher in Mayurbhanj (see Table 2.16).

Table 2.13 Size group wise distribution of number of operational holdings in study districts of Orissa

	Size groups (area in hectare)					
District	Marginal	Small	Semi-medium	Medium	Large	All sizes (total)
Keonjhar	124,703	81,364	31,290	11,704	4105	253,167
Mayurbhanj	173,331	106,723	54,586	18,665	4322	357,627
Sate total	1,921,842	1,497,752	918,947	381,272	132,201	4,852,014

Source Agricultural census 2010–11

Table 2.14 Agro-climatic characterisation and classification in study districts of Orissa

District	Annual rainfall (mm)	Climate (based on moisture index)	Net irrigated area (%)
Keonjhar	1484	Dry subhumid	25
Mayurbhanj	1553	Moist subhumid	29

Source ICAR-Central Research Institute for Dryland Agriculture

Table 2.15 District-wise cropping intensity in study districts of Orissa, 2011–12

District	Net sown area	Gross cropped area	Cropping intensity (%)
Keonjhar	259	395	152
Mayurbhanj	335	479	143
State total	5292	8801	166

Area in '000 hectare

Source Directorate of Agriculture & Food Production, Odisha (2011–12)

Table 2.16 Irrigation status in study districts of Orissa

District	Net irrigated area	Gross irrigated area	Total cropped area
Keonjhar	74.37	106.52	371.67
Mayurbhanj	112.78	164.03	449.5
Total	2078.9	3087.18	8267.75

Source Directorate of Agriculture & Food Production, Odisha (2011–12)

2.6 Conclusion

The aim of this chapter was to provide an overview of rice economy and the performance of India in terms of rice trade. This chapter also discussed the trends in rice production, area under cultivation and yield for rice in the states selected for the purpose of analysis. A detailed district-wise analysis of cropping pattern, irrigation status, etc., is also discussed. The next chapter will discuss the major rice production technologies and the need for sustainable rice production practices to ensure rice security and food security.

Chapter 3
The Need for Sustainable Rice Cultivation Practices

3.1 Introduction

The present chapter will dwell on the major rice production technologies by discussing the importance of sustainable agricultural practices in general and rice cultivation in particular. Section 3.2 of this chapter will talk about the major rice production technologies, and Sect. 3.3 will highlight the need for sustainable agricultural practices. The importance of system of crop intensification in general and the SRI in particular are discussed in detail in Sects. 3.4 and 3.5.

3.2 Rice Production Technologies

The Green Revolution that was initiated in the 1960s had a profound impact on the Indian rice economy. Since 1960s, India started partnership with International Rice Research Institute (IRRI). This led to the introduction of IR8 high-yielding rice variety. Subsequently, more than 400 improved rice varieties with resistance to pests and diseases were also introduced. This helped in streamlining rice production practices and extensive information exchange with Indian rice scientists (Revilla-Molina et al. 2012). In addition, there was commercialisation of hybrid rice in India due to the partnership with IRRI.

Yamano et al. (2016) has provided a detailed review of the new rice technologies that have been developed by international agricultural research centres in collaboration with their partners. The focus of the review was on technologies that have been already adopted by the farmers.

3.2.1 Adoption and Impact of Improved Germplasm

Ever since the introduction of modern variety IR8 that was made available to millions of farmers in Asia and Africa, various improvements in the rice varieties

P. Varma, *Rice Productivity and Food Security in India*,
DOI 10.1007/978-981-10-3692-7_3

have taken place to enable countries to make rice accessible and affordable. This helped in achieving the objectives of food security. Along with the development, adoption and diffusion of modern rice varieties, there have been plenty of studies analysing the impact of the adoption of modern rice varieties in general (Khanal and Maharjan 2014; Tiongco and Hossain 2016) and for India in particular (Singh and Kumar 2014; Bansal and Ward 2015).

A lower growth rate in rice yield during the late 1990s and the early 2000s raised questions about the effectiveness of rice breeding programmes. In South Asia, farmers continuously use older modern rice varieties that were released more than several decades ago (Pandey et al. 2015). For example, Swarna, which was released in 1979, is planted on 4.3 million ha or more than 30% of the total rice area in the three eastern Indian states of West Bengal, Odisha, and Chhattisgarh. In Bangladesh, two varieties dominate the Boro season: BRRI dhan28 (released in 1994, accounting for 19% of rice area) and BRRI dhan29 (released in 1994, planted on 14% of rice area). During the main rice season, Aman, BR11 (released in 1980), occupied 14% and Swarna occupied 6%. The popularity of these old varieties is despite the release of newer modern rice varieties that are supposedly superior in terms of yield, resistance to pests, or tolerance of abiotic stresses.

Stress-Tolerant Rice Varieties

Environmental stresses such as drought or flood place severe constraint in rice production and affect millions of poor in Asia alone (Dar et al. 2014). Flooding is a frequent hazard in low land regions and is expected to intensify with climate change. Short-term submergence regularly affects over 20 million hectares of rice. Therefore, rice productivity in stress-prone areas of India is not only low but also very fragile. The feasible approach to solve this problem is to develop rice cultivars with tolerance of submergence and with agronomic and quality traits acceptable to farmers. Stress-tolerant high-yielding rice varieties developed and tested for specific agro-climatic ecosystems. Stress-tolerant improved rice varieties were introduced to reduce losses in rice production under stress conditions. The distribution of stress-tolerant rice varieties in South Asia started in 2008 and was implemented under the project Stress-Tolerant Rice for Africa and South Asia (STRASA) (Yamano et al. 2016). Stress-tolerant rice varieties are tolerant of abiotic stresses such as submergence, drought, flood, and salinity (Yamano et al. 2016). They are suitable for specific agro-climatic ecosystems. For example, Swarna Sub1 and Varshadhan rice varieties give high yields under flood conditions, while Sahbhagi dhan cultivar (IR 74371-70-1-1) is drought-tolerant (Yamano et al. 2013). IRRI and its collaborators developed Swarna-Sub1 which is a submergence-tolerant rice variety (Yamano et al. 2013). In order to promote the adoption of new high-yielding stress-tolerant rice varieties by the farmers and also to make the seed available locally, the National Food Security Mission (NFSM) allocated thousand ha cluster demonstrations to IRRI for Kharif 2012–2013 (IRRI 2012–2013). The important submergence-tolerant rice varieties include Swarna-Sub1, Samba Mahsuri-Sub1, and IR64-Sub1 in India; BBRI dhan52 (BR11-Sub1) and BRRI dhan51 (Swarna-Sub1) in Bangladesh; and Samba Mahsuri-Sub1 and Swarna-Sub1 in Nepal (Yamano et al. 2016). Similarly, drought-tolerant variety Sahbaghi dhan was also released recently in India, Bangladesh (with the name of BRRI dhan56) and Nepal.

3.2.2 Adoption of Natural Resource Management Practices

Mechanized Direct Seeded Rice (DSR) Technology for Rice Cultivation

The transplanting of seedlings into puddled soil is the common practice while cultivating rice. This provides the benefit by reducing water percolations losses, controlling weeds and facilitating easy seedlings (Kumar and Ladha 2011). However, repeated puddling can have an adverse impact on the physical properties of soil by destroying soil aggregates, reducing permeability in subsurface layers, and forming hard-pans at shallow depths, all of which can negatively affect the following non-rice upland crop in rotation (Kumar and Ladha 2011). Additionally, puddling and transplanting is input-intensive as it requires large amount of water and labour. Since water and labour are becoming increasingly scarce, expensive rice production is also becoming less economical and sustainable. As a result, a shift away from puddled-transplanted rice production to direct seeding of rice is necessary. Direct seeding helps in saving the cost of cultivation by conserving labour and water. It also enables timely sowing that helps achieve higher yields for rice, as well as for the succeeding winter crop (Wheat) (Pathak et al. 2011). The crop matures earlier than the traditional practice by 7–10 days (Joshi et al. 2016).

Alternate Wetting and Drying (AWD) Irrigation

Alternate wetting and drying (AWD) can help in increasing rice yield by reducing water requirement. AWD is one of the water-saving technologies that have been developed to encourage farmers to use water efficiently. It was developed in the 1990s by the IRRI (Howell et al. 2015). The AWD allows the soil to dry out before re-irrigation (Howell et al. 2015). The partial drainage of rice fields which is part of AWD is done by irrigating the fields to the desired depth and then re-irrigating after some time (Lampayan et al. 2015). This method can reduce the water use by the framers. Despite these benefits, the adoption rates of AWD are limited in many rice-producing regions (Howell et al. 2015). The modified method known as Safe AWD has 3 key elements. They are (1) shallow flooding for the first two weeks after transplanting; (2) shallow ponds from heading to the end of flowering. (This is a crucial stage when the rice crop is very sensitive to water-deficit stress) and a time when the crop has a high growth rate and water requirement; and (3) irrigation water is kept applied whenever the perched water table falls to about 15 cm below the soil surface during all other periods (Yamano et al. 2016).

3.3 The Need for Sustainable Practices

The food production in India improved tremendously due to the Green Revolution and thereby reduced the food insecurity and poverty to a great extent (Bishwajit et al. 2013). Despite the progress achieved during this phase, millions of people still remain in food insecurity and poverty. Although the Green Revolution was successful in dramatically increasing food production, input-intensive farming has produced severe environmental concerns. Additionally, the increased yields achieved during the Green Revolution period have not been maintained in the recent years. For example, the yields of major cereals, including rice, have not increased at the pace recorded from the 1950s to the 1990s (Yamano et al. 2016). Therefore, achieving self-sufficiency in rice production and improving the rice yields through sustainable farming practices are very crucial to ensure national food security.

Since the 1990s, India's agricultural growth has stagnated and remains at less than 2%. This is much lower than the growth rates of other sectors. This poses a serious threat in terms of meeting the food demands of a large and growing population. The main challenge is to meet the growing food demand in a more sustainable manner without harming the environment. During 2004–2008, the average estimated growth in foodgrain production was 1.98%. This was higher than the average population growth rate of 1.5% during the same period. In 2006, the contribution of agricultural sector to national GDP was only 16%. However, around 70% of India's poor, who mostly live in rural areas, depended on agriculture for their livelihoods (Gujja and Thiyagarajan 2009).

Another major issue that the country is facing is severe water scarcity. Around 45% of all the available water in India is used for irrigation purposes. Groundwater constitutes around 70% of water used for irrigation purposes. As per the study done by World Bank, by 2020, the demand for water by India will exceed all sources of supply (Gujja and Thiyagarajan 2009). As a result, the situation necessitates the development of better agricultural and water management practices to address the emerging challenges. In order to face the challenge of decreased water availability, it is important to explore sustainable agricultural practices. This is important to revive and sustain agricultural productivity to feed the millions of population. Another set of challenges emerges from the injudicious use of fertilizers and pesticides, water pollution, indirect damage to ecosystems, etc. (World Bank 2008).

The population of India is projected to reach around 1.59 billion by 2050. This is approximately 470 million people more than what we have today. If the population is going to grow at this pace, India might become the world's most populous country by 2035. Ironically, even today there are around 200 million underfed people and 50 million on the verge of starvation. Compared to 1997, today the number of hungry people has increased and we are importing food at great expense. All these pose serious challenges to the food security situation in India (Gujja and Thiyagarajan 2009).

The constraints that the country is facing vary from state to state and also from region to region. The eastern region cultivates a major portion of rice in India. This zone has high rainfall, drought, and severe floods, which makes the rice production very vulnerable (CRRI 2011). In certain areas, the farmers lack suitable high-yielding varieties and quality seeds, while in some other areas, the types of soil affect the productivity (CRRI 2011). Some of the major constraints faced by the regions are summarised in Table 3.1.

As mentioned earlier, the average rice productivity in India is much lower than the global average (CRRI 2011). The rice productivity is also lower than that of other major rice-producing nations in South Asia such as Bangladesh and Sri Lanka. The projections of population growth, rice production, and demand are given in Table 3.2.

Table 3.1 Major constraints faced by rice-producing regions in India

Constraints	States/agro-ecologies
Small and marginal farmers with poor resources to use optimum/recommended inputs	Mostly Eastern states of India
Erratic rainfall with poor soils	Madhya Pradesh, Odisha and some parts of Uttar Pradesh
Flash floods, water logging due to poor drainage	Assam, West Bengal, North Bihar and Eastern Uttar Pradesh
Use of traditional varieties	Mostly Eastern states
Low and imbalanced use of fertilizers	North-Eastern and Eastern states
Delayed and prolonged transplanting due to delay in monsoon	Mostly rain-fed lowlands
Poor adoption of production technology	Mostly in uplands and lowlands
Saline and alkali soils	West Bengal, Odisha, Andhra Pradesh, Tamil Nadu, Kerala, Karnataka, Maharashtra, Gujarat, Western Uttar Pradesh, Punjab, Haryana, etc.

Source CRRI (2011)

Table 3.2 Rice requirements at various levels of per capita availability and projection at different production growth rates

Year	Population (million)	Projected demand (mt)	Projected supply (mt) at growth rates of		Demand–supply gap (mt) at growth rates of	
			1.34%	1.14%	1.34%	1.14%
2005	1116	94.5	91.8[a]	91.8[a]	2.7	2.7
2010	1210	100.7	94.0[a]	94.0[a]	6.7	6.7
2015	1288	106.8	101.5	100.5	5.3	6.3
2020	1370	112.8	108.5	106.4	4.3	6.4
2025	1445	117.3	116	112.5	1.3	4.8
2030	1523	121.6	123.9	119.1	−2.3	2.5

Note [a]Actual figure for the year
Source CRRI (2011)

3.4 The Need for Sustainable Intensification of Agricultural Practices—System of Crop Intensification (SCI)

In order to meet the global food security requirements throughout the twenty-first century, agricultural sectors around the world will need to pursue strategies for sustainable intensification of agricultural production. Although the terminology used can vary—low-input intensification, sustainable crop production intensification, or sustainable agricultural intensification—the intended redirection of thinking and approach is more or less the same (SRI International Network and Resources Center 2014). An underlying feature of these recommended changes is their divergence from the kind of agricultural strategy that had prevailed over the past 50 years. Technologies for "modern agriculture", particularly those associated with the Green Revolution, have been instrumental in increasing production by encouraging farmers to use more land, water, machinery and inputs (SRI International Network and Resources Center 2014). However, these input-intensive methods of production have led to several concerns being raised about their impacts on environment and sustainability.

An alternative strategy for intensification that can be broadly characterised as agroecological aims to make the most productive use of the available limited resources. There is growing scarcity of land and water resources on one side and growing demand due to the population pressures on the other side. Additionally, the quality of the resources is also degrading and getting polluted. The increasing scarcity of our natural resources vis-à-vis the needs of our growing populations places an ever-greater pressure on the need to improve the management of the soil systems, water and biotic resources (SRI International Network and Resources Center 2014).

Since virtually all the land is cultivated and resources such as water is severely constrained, any additional production of rice to ensure food security needs to come from improving rice productivity through sustainable practices.

3.5 The Case of SRI

Rice cultivation in India is heavily dependent upon rainfall. Water scarcity is one of the major threats that the rice cultivation facing today. Inadequate rainfall, lack of water harvesting measures and unscientific use of water for agriculture are the main reasons for decline in per capita availability of water by 40–60% in many Asian countries including India. As a result, several farmers are forces to abandon rice cultivation. Since rice is heavily dependent on water, identification of alternative rice cultivation strategies that require less water is the need of an hour. To mitigate the looming water crisis, there is an urgent need to develop water-saving technologies and production system that increases or sustains rice production.

Although the exact amount of food required to feed the growing population may be debatable, undoubtedly India will have to produce a lot more paddy to meet the growing demand for rice. As per some estimates, the amount of milled rice required by 2030 is roughly around 130 million tonnes (Gujja and Thiyagarajan 2009). Unless we focus on measures to enhance rice yields without harming the environment, we will not be in a position to achieve this target. Unfortunately, we are witnessing an increased shrinking of area under rice cultivation and as a result fluctuations in annual production and stagnating yield. All these are contributing to an escalating input cost as well. The cost of cultivation of paddy has consistently been increasing owing to the increased costs of seeds, fertilisers, labour and other inputs. Labour availability is also a cause for concern. As a result of all these, the challenge of sustaining farmers' interest in rice cultivation is also increasing. There is clearly an urgent need to find ways to grow more rice, but with less water and fewer inputs. There have been efforts to develop technologies to reduce water use for rice cultivation. AWD method is one such technology. But these methods rarely showed much promise in terms of increasing rice farmers' net income. It is in this context that SRI has been developed, which not only helps in reducing the amount of water but also helps in increasing rice yields substantially. Thus, SRI is widely believed to enhance the livelihoods and overall welfare of rice farmers. Besides increased paddy production, the SRI contributes to the enormous savings of water and seed and other resources which are increasingly getting scarce day by day. The input requirements under both traditional and SRI practices, assuming that SRI is practised on 20 million hectares of rice (out of India's current 43 million ha under rice), are summarised in Table 3.3 (Gujja and Thiyagarajan 2009).

SRI does not require any specific crop variety. High-yielding varieties or traditional ones, hybrids or landraces, all perform better when planted with SRI methods, although some varieties respond better than others. Generally, farmers show greater interest in growing local varieties while applying SRI methods. This often helps then in easy adaptation to local soil and climate conditions, and better pest and disease resistance. Additionally, the demand is also higher for local varieties. SRI methods allow farmers to maintain and increase the diversity of planted rice varieties, something that has been diminishing in recent years.

Table 3.3 Impact of SRI in India if adopted on 20 million hectares

	Level		Total estimate		Advantage due to SRI
	Current	SRI	Current	SRI	
Seed use	30 kg/ha	7.5 kg/ha	600 million tonnes	150 million tonnes	450 million tonnes saved
Irrigation water	149 m^3	92 m^3	2980 million cubic metre	1840 million cubic metre	1140 million cubic metre saved
Paddy production	3.17 tons/ha	4.17 tons/ha	139 million tonnes	183 million tonnes	44 million tonnes extra production

Source Gujja and Thiyagarajan (2009)

Thus, SRI is evolved a powerful alternative to help rice farmers to overcome the present water crisis. Additionally, it helps farmers in achieving higher productivity by altering the plant, soil, water and other management practices.

The SRI methods do not leave behind the ecological footprints as in the case of conventional paddy production practices. The ecological footprints are due to the overuse of fossil fuels and synthetic fertilizers that pollute freshwater supply (Bhatt 2015). Excessive use of water contributes to greenhouse gases emission and, thereby, more global warming. In addition to this, the cost of production, especially the cost of inputs, is high under the conventional practices and the use of synthetic fertiliser to increase production is resulting in diminishing or constant returns to scale (Bhatt 2015). Therefore, input management practices such as SRI are very essential in improving the yield for ensuring food security of the country.

SRI is not a technology, and hence, it has been purposefully called a system rather than a technology. SRI involves a holistic management practice to provide ideal growing condition for rice plant. The next chapter will discuss the salient features of SRI in greater detail.

Chapter 4
The System of Rice Intensification (SRI)

4.1 Introduction

This chapter looks at SRI in greater detail to enhance our understanding. Section 4.1 will cover the salient features of SRI. Section 4.2 will talk about the wider applications of SRI principles to other cropping systems. Sections 4.3–4.7 will talk about the experience of SRI at the international, Indian and the selected district level. Section 4.8 will discuss the gender implications of SRI adoption, and Sect. 4.9 will briefly discuss the key actors and the role of institutions in the spread and adoption of SRI. Conclusions are given in Sect. 4.10.

The SRI originated in Madagascar during the 1970s and 1980s, and the father of this invention is credited to the French priest Henri de Laulanie. Improving the livelihoods of the poor and small-scale Madagascar farmers was the major objective behind testing the unusual rice farming practices. In 1994, seeing the success of the SRI practices by Cornell International Institute for Food Agriculture and Development (CIIFAD) that worked with Henri de Laulanie and his Non-Governmental Organisation (NGO) called Association Tefy Saina, Norman Uphoff, CIIFAD Director from 1990 to 2005, supported the spread of SRI from Madagascar around the world (Dill et al. 2013). Thus, the SRI is widely recognised as a promising systemic approach to rice production for small-scale producers. Unlike treating it as conventional agricultural technology, it is considered as a set of ideas and insights by modifying agronomic practices for improving the rice productivity (Uphoff 2016). To generate higher yield output, it has been shown that the SRI technique does not depend on the use of improved or new varieties or on the use of synthetic fertilisers and agrochemical crop protection. SRI studies have highlighted that the adoption of SRI is environment-friendly, enhancing yield and substantially reducing rice's dependence on water and other input requirements (Stoop et al. 2002; Uphoff 2002, 2003) by changing the management of plants, soil, water and nutrients (Satyanarayana et al. 2007). Unlike most current agricultural technology, SRI is not a technology but rather a system or a methodology (Uphoff

P. Varma, *Rice Productivity and Food Security in India*,
DOI 10.1007/978-981-10-3692-7_4

2016). Therefore, farmers are more of adapters rather than simply adopters (Fig. 4.1).

The SRI method is a complete deviation from the conventional way of paddy cultivation. SRI is a low external input method and, in sharp contrast to the Green Revolution experience, it increases yield with lower inputs. This is achieved mainly by focusing on the soil rather than the plant, and the promotion of organic fertilisers (like compost) will allow the deviation from the application of chemical fertilisers (Dill et al. 2013). As a result, the healthy soil will provide the right environment and nutrients for growth of the roots of the plant resulting in a strong and fecund plant. Thus, the maximum potential of the plant is being utilised. This is because the plants can better express their genetic potential when grown in a more optimal environment.

Unlike other agricultural technology, SRI is not a package of fixed technical specifications. Rather it is a system of production based on certain core principles of plant soil and water management. This definition of SRI necessitates the adaptation of the system to specific conditions of an area rather than trying to impose practices relevant to one location on the other injudiciously (Namara et al. 2003). These characteristics make SRI distinct from the standard technological innovations. The comparison of SRI with traditional paddy/rice cultivation systems is shown in Fig. 4.2. The following summarises the SRI components of five to six major practices:

1. Transplanting of young seedlings
2. Shallow planting of seedlings
3. Single seedling at wider spacing
4. Weeding by mechanical weeder

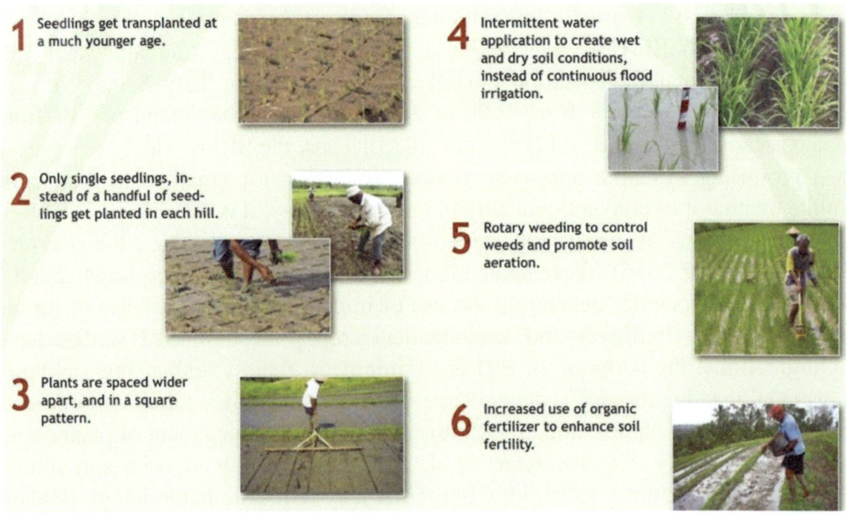

Fig. 4.1 Major SRI practices. *Source* World Bank Institute

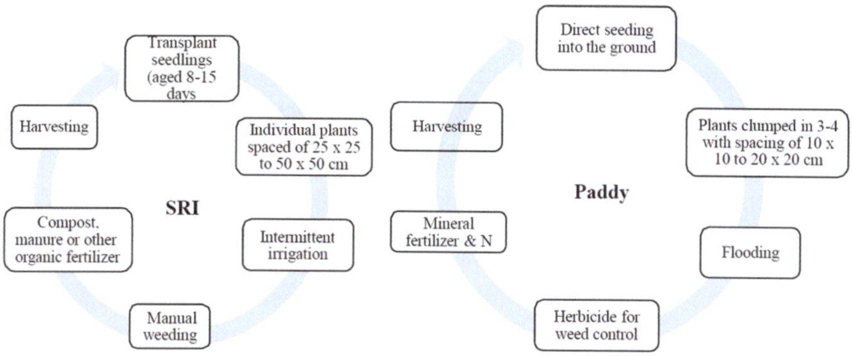

Fig. 4.2 SRI in comparison with traditional paddy rice cultivation systems. *Source* Aune et al. (2014)

5. Use of organics
6. Efficient water management: Alternate wetting and drying

Transplanting of young seedlings: Seedlings that are 8–12 days old, but less than 15 days old.

Shallow planting of seedlings: To be planted in an un-flooded nursery and then removed gently, with minimum shock to their roots, being replanted in the main field carefully, quickly and shallow (1–2 cm).

Single seedling at wider spacing: Plants are planted singly instead of 3–6 plants, and in a square pattern, usually 25 × 25 cm. Plant densities in the field are reduced by 70–90%. This gives plants' roots and canopies more room to grow and spread, acquiring more nutrients and sunlight (Uphoff 2016).

Weeding by mechanical weeder: This promotes aeration of the soil better than possible with hand weeding or use of herbicides. Active soil aeration can stimulate paddy yields by 1–2 tons/ha (Uphoff 2016).

Use of organics: SRI emphasises the enhancement of the soil's organic matter by adding compost or other biomass. This has been shown to improve the soil's structure and functioning as well as provide the plants with more complete and balanced nutrition (Uphoff 2016).

Efficient water management: Alternate wetting and drying: Although SRI requires less water, the management of water is a very essential component. Therefore, the field should be intermittently wetted and dried, so that the soil is mostly aerobic, never hypoxic. Good drainage of the soil is almost as important as the provision of irrigation water. A lack of oxygen in the soil will suffocate the plants' roots as well as the aerobic soil organisms that can provide many beneficial services to the plants (Uphoff 2016).

The above practices of SRI can be grouped under three main principles—soil management, water management and plant management.

Soil management: The soil quality is improved through the use of organic matter. Regular intervals of weeding will aerate soil, stimulate soil biota and enhance the nutrient fixation in the soil. This is effectively done by using a mechanical rotary weeder.

Plant management: In order to provide ideal space and conditions, shallow planting of young and single seedlings is required, and this enhances their potential for root development and tillering. Ideally, the seedlings are 8–12 days old, and these seedlings are planted gently and horizontally into the soil. The distance between each seedling should be ideally at least 20 cm.

Water management: In order to follow and ideal water management in the paddy field, the fields must be moist and not flooded during the plants' vegetative growth phase. This should be ensured until the stage of flowering and grain production. Although the amount of water required is less, SRI necessitates proper management of water.

4.2 System of Crop Intensification—Wider Application of SRI Principles

There are many innovations that are taking place within the conceptual and practical framework of the adoption of SRI. The experience of farmers and professionals from several countries shows that they are experimenting with the concepts and methods of SRI into a variety of other crops. Additionally, we can see that the principles of SRI that was first applied to irrigated rice production are not being extended to rain-fed rice and then to improving yields of a variety of other crops (Uphoff 2012). The SRI is thus classified as system of crop intensification (SCI) due to its wider scope and application. The broad applications of SRI principles can be seen in other cereals, legumes and vegetables (Abraham et al. 2014). System of wheat intensification (SWI), system of teff intensification (STI), system of finger millet intensification (SFMI) and sustainable sugar initiative (SSI), etc. are examples of such wider application of SRI principles to other crops. The principles of SCI and SRI, based on demonstrated agronomic practices, are shared with other agroecological domains of innovation such as agroforestry, conservation agriculture, integrated pest management, and integrated range and livestock management.

The SCI approach is not just about deriving more output from a given levels of input rather it is about achieving higher levels of output with less use of resources such as land, labour, capital and water. This is achieved by making modifications in crop management practices. The growth and spread of SRI and SCI in various countries are depicted in Fig. 4.3.

The SCI is more about plant management rather than pest management. SCI adopts a four-pronged strategy which is implemented systematically especially in the case of soils which is of poor quality. The approach involves soil preparation and management, crop spacing, systematic application of locally prepared organic inputs and micronutrient foliar sprays.

As summarised by Abraham et al. (2014), the common elements involved in SCI crop management are the following:

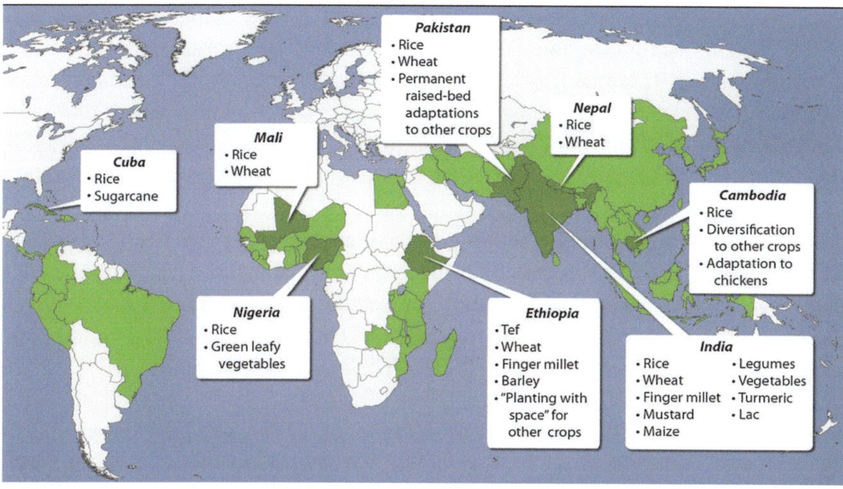

Fig. 4.3 The widespread dissemination of SRI and SCI ideas and practices. *Note* In the *light green* coloured countries, SRI methods have been seen to produce better phenotypes from available rice genotypes; in the *dark green* coloured countries, in addition to this, there has been experimentation with and confirmation of SCI principles and techniques; lists for each country show the crops which to date have been shown to improve yields with SCI methods. *Source* SRI International Network and Resources Center (2014)

- Establishment of healthy plants provides conservation and nourishment for root system growth and for the associated shoot growth.
- Transplanting or sowing plants with wider spacing to significantly reduce the plant density. This provides each plant with more room to grow both above and below the ground.
- Enrichment of the soil with organic matter: This is done by keeping the soil well-aerated to support the better growth of roots and of beneficial soil biota.
- Appropriate application of water ensures a favourable growth of the root and soil-microbial growth. This helps in avoiding hypoxic soil conditions that adversely affect roots and aerobic soil organisms.

Transplant of young seedlings—one of the key components of SRI—has been shown to have positive effects on other crops such as finger millet, mustard and teff (Abraham et al. 2014).

4.3 Pros and Cons of SRI

There have been plenty of debates surrounding the adoption of SRI. The opponents of SRI argue that SRI lacks scientific rigour and accuracy of measurement (Dill et al. 2013). Since it does not follow a fixed practice and has enormous flexibility in the manner in which a farmer and a region adapt SRI, comparison of SRI adoption and its benefits across regions becomes an unenviable proposition. One major criticism against SRI is that it is labour intensive and therefore not suitable to many regions.

Another line of criticism is with respect to the higher yield benefits reported by SRI. Majority of articles in this context were published in the mid-2000s (Dobermann 2004; Sheehy et al. 2004; Sinclair and Cassman 2004; McDonald et al. 2008). The scepticism was surrounding the extraordinary high yields, effectiveness of SRI practices and the experimental procedures (Chapagain et al. 2011). On the other hand, the proponents of SRI highlight several benefits of SRI which, among others, include higher yields and greater water productivity (Stoop et al. 2002; Uphoff 2007; V&A programme 2009; Thakur 2010; Uphoff et al. 2010; Geethalakshmi et al. 2011; Palanisami et al. 2011, 2013; Chowdhury et al. 2012). Additionally, they stressed upon the environmental benefits of SRI on account of reduced input requirements such as water and fertiliser. However, one of the recent studies by Gathorne-Hardy et al. (2016) argues that SRI gains in terms of environmental and economic benefits which is at the expense of employment. Despite all these criticisms, SRI methods are gaining popularity, and there is statistical evidence that the practice of SRI has steadily increased among the farmers; in the year 2013, SRI methods have been validated in more than 54 countries (Styger 2014). The available information shows that more than 10 million farmers benefit from SRI (Styger 2014).

SRI has gained popularity as one of the climate-smart agriculture (CSA) measures and has garnered support by institutions ranging from farmers' organisations to non-government organisations and the World Bank. In 51 countries, the SRI practices have been validated with their governments planning to expand the application of SRI (Gujja and Uphoff 2013), though its popularity and controversy have been the background of its debate. SRI has been a subject of discussion among the scientific community, and there have been studies cited as showing similar yields for SRI as compared to other recommended management practices (Aune 2014). The summary of the most important arguments of the proponents and opponents of SRI is given in Table 4.1.

4.4 International Experience of SRI

Ever since the SRI was originated in Madagascar in the 1980s, there has been a rapid spread of the SRI to a number of other countries, particularly in the major rice-growing nations of China, India, Indonesia, Vietnam and Cambodia, which produce two-thirds of the world's rice (see Fig. 4.4). There has been a steady increase in farmers practicing some or all SRI principles, and in the year 2013, SRI methods have been validated in more than 54 countries with support from several governments objective to expand SRI (Styger 2014). The available information shows that more than 10 million farmers benefit from SRI (Styger 2014). Studies

Table 4.1 Pros and cons of SRI

Aspect	Arguments of proponents	Arguments of opponents
Economic	**Higher yields**: The increase in yield was in the range of 20–200% of conventional rice farming yields	**Difficulties in proving higher yields**: The higher yields reported in one region is difficult to replicate in another region, partly because SRI is an adaptive methodology rather than a technology
	Lower production costs: SRI uses less seeds (up to 90% less), less water (25–50% less) and less pesticides and chemical fertilisers (both up to 100% less). Although SRI can be labour intensive in the initial years but later it can become labour-neutral or even labour-saving. Total input costs are reduced	**Labour intensivity**: SRI is more labour intensive and therefore is only suitable for small land sizes
	Reduced risk of crop failure: SRI produces robust plants with strong tillers and healthy root systems. The crop is more resilient to pests and diseases and more robust under extreme temperatures, storms and droughts which are increasing in the context of climate change	**Increased risk**: Transplanting single, very young seedlings bears a high risk of snails, crabs and rats eating the plants. Also, heavy rainfall easily destroys the transplanted seedlings
	Higher prices: SRI generates higher quality and thus likely to receive a premium price. For example, to increase small-scale farmer's profits and contribute to food security, SRI can be sold as more expensive seed-rice with generation of higher yields, less inputs, fewer crop failures and higher prices	
Environmental	**Market opportunities**: Demand is strong and growing for agricultural production methods that produce chemically free foods creating human health benefits also enhancing the quality of soil and water affected	**Adoption**: If farmers do not adopt SRI easily, it may not be beneficial for them.
	Better soil quality: Practicing SRI results in a greater abundance, activity and diversity of soil organisms, and thereby improves its quality	**Organic matter**: There will not be enough organic matter available to practice SRI on a large scale
	Prevention of water pollution: SRI practice is known for reducing the pollution of ground water from rice farming	

(continued)

Table 4.1 (continued)

Aspect	Arguments of proponents	Arguments of opponents
	Natural resources: SRI protects the nature by saving water consumption by rice farming and promotes organic fertilisers rather than the production of chemical fertilisers which relies on oil and other natural resources	
	Climate change mitigation: SRI plots are likely to emit lower methane gas emissions than traditional plots	**Climate change**: SRI plots have shown to emit more nitrous oxide than traditional rice plots, resulting in adverse effects on climate change
	Agro-biodiversity: SRI promotes diversification of soil biota, animals and plants in and around the paddy field, owing to the lower use of agrochemical inputs. Since SRI works with all types of rice varieties, this contributes to maintaining a diversity of rice varieties	**Varieties**: High-yielding varieties are necessary to feed the growing world population
Social	**Social empowerment**: Encouraging farmers to experiment and to engage in participatory technology development, hence, building up adaptive capacities	
	Positive impacts on human health: Improved water quality and less physical contact with chemicals	
Methodology	**Upscaling**: SRI has a high potential to upscale because of its application to a variety of areas, on different scales and even with different types of crops. However, upscaling requires proactive farmers, motivated extension staff and convincing political support	**Difficulties to evaluate SRI scientifically**: SRI is not standardised. There is no uniform definition as the principles can be applied partially and flexibly. Hence, making the understanding of SRI too esoteric and causing difficulty in its evaluation. It is analogous to what is known as "best management practices" (BMP) **Dissemination**: Farmers prefer clear recommendations to follow

Source Dill et al. (2013)

with research and demonstration plots in several tropical countries have shown that in comparison to traditional rice farming practices, SRI methods generate productivity and contribute to being resource-saving and environmentally friendly (Bassey 2016). It is reported that, as compared to traditional rice plants, SRI plants have greater resistance to biotic and abiotic stresses (Uphoff 2010).

World rice production has a huge diversity in terms of relative access to land, labour and capital, as well as their adaptation to very diverse agroecological

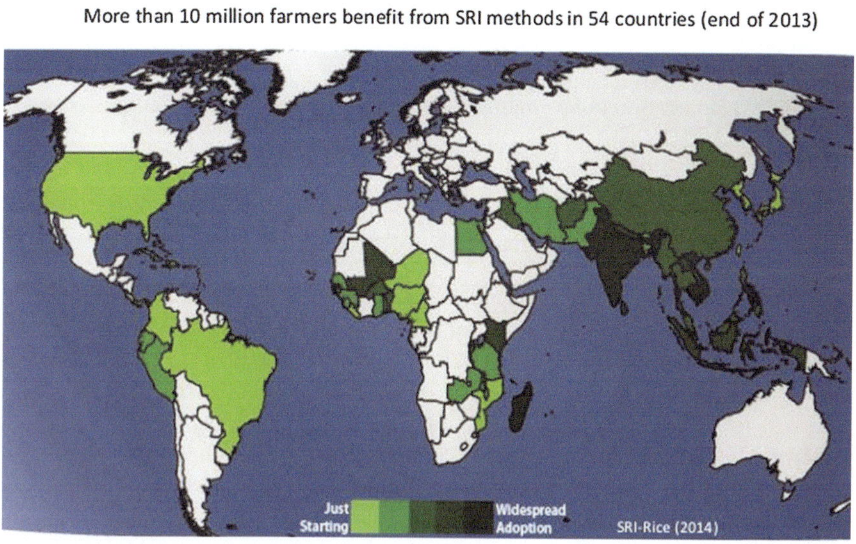

Fig. 4.4 Spread and adoption of SRI. *Source* Styger (2014)

environments. Therefore, SRI can never be introduced as a standardised package. As a result, partial adoption of SRI is very common, and the contributors influencing the adoption of SRI, the constraints in adoption and the outcomes of adoption vary quite considerably across countries and regions. The review of region-specific studies on SRI undertaken below is an eloquent testimony of this fact.

Asia: Experience of SRI in Asia was quite significant, and this is not surprising considering the fact that around 90% of the world's rice is produced in this region SRI took roots in all rice-producing countries of South, Southeast and East Asia. Malaysia, the Republic of Korea and the Democratic People's Republic of Korea (DPRK) started the evaluation and adoption a little later than other countries (Uphoff and Kassam 2009). During 2008, Norman Uphoff of Cornell University was invited by a group of professionals to visit Malaysia to discuss SRI with the Minister of Agriculture and others interested in the rice sector. Subsequently, he had a meeting with the rice scientists at Malaysian Agricultural and Rural Development Institute (MARDI), civil society representatives and the faculty of the National University of Malaysia (UKM). Following this visit, Dr. Anizan Isahak at UKM took the initiative to form an SRI research group at UKM called "SRI-PADI".

As far as the Republic of Korea is concerned, the first trials of SRI methods were undertaken in 2007 by Dr. Jinyong Choi, professor of sustainable agriculture at Gyeongsang National University (GNU). In 2011, two related articles on planting densities and nutrient uptake suggested that SRI methods are useful for small-scale rice producers in Korea. The articles have also pointed out the sustainability nature of the SRI methods.

In 1998, a delegation of few members from the Academy of Agricultural Sciences (AAS) from DPRK made a visit to the Cornell University to discuss about SRI. Subsequently, in 2005, a larger delegation including AAS scientists visited Cornell University under the auspices of The Asia Foundation. This meeting resulted in a long discussion on SRI benefits. The first experiments with SRI methods were undertaken in 2008 with the support from four cooperatives in the Pyongyang area and the Ministry of Agriculture, under the auspices of the Korean Committee for Solidarity. The trials showed a yield increase of 0.5–1.0 tons/ha by using just some of the practices adapted to local conditions. By 2009, the area under different versions of SRI practices by the cooperatives increased to 250 ha. In 2010, trials were underway to assess what would be optimally wide spacing under local conditions.

Africa: Except for Madagascar, the country of SRI origin, there has been a delay in spreading the SRI methods in this region (Uphoff and Kassam 2009). The trade imbalances and the excess demand for rice prompted the region to look for ways to increase rice production (Styger and Jenkins 2014). Farmers from Mali to Benin, Senegal, Nigeria and Sierra Leone, among others, have taken the initiative for adoption of SRI based on trials in small plots to regional level trainings and initiatives (Styger and Jenkins 2014). Unlike South and Southeast Asia, where wide expanses of irrigated rice fields predominate, West Africa consists of a broad array of different rice production systems, the largest of which is rain-fed upland. The experience till now shows that SRI has grown more quickly around the world in areas of irrigated rice production, as it requires less adaptation (Styger and Jenkins 2014).

Latin America: Cuba was the first country in this region to adopt the SRI methods. In Cuba, the method was known as Sistema Intensivo de Cultivo Arrocero (SICA) (Uphoff and Kassam 2009). The validity of SRI practices has also been demonstrated in countries like Brazil, Ecuador, Peru and Costa Rica. The initial inertia in adoption in this region seems to be the reluctance of rice farmers in Latin America to transplant rice by hand, favouring mechanised methods, given the costs and availability of agricultural labour (Uphoff and Kassam 2009).

Middle East and North Africa: The rice research institutes in Iraq, Iran and Egypt have all demonstrated the efficacy of SRI management methods. The first on-farm trials were conducted in the Baghlan Province of Afghanistan in 2007 by the Aga Khan Foundation with EU funding. The trials were undertaken with vigorous tillering but insufficient grain filling. As per the rice yield is concerned, the SRI trial plots gave only 3 tons/ha compared with 6.5 tons/ha using usual methods. However, the planting of SRI methods was done one month later than normal, and at an altitude of 1700 m elevation, where the growing season is short. Fortunately, farmers were impressed by the tillering they noticed and as a result they were not deterred by the initial results.

The country-level experiences show that the increased yield attained with less water in SRI method was the major attraction for countries like China, India and Egypt (Stoop et al. 2002). Owing to the labour-saving feature of SRI methods, many farmers in countries like India and China have been attracted to this technique (Uphoff et al. 2011). Nonetheless, some of the early studies have pointed out that SRI is inherently labour intensive and therefore not suitable to regions with labour constraints (Moser and Barrett 2003). A later study pointed out that SRI will reduce the labour requirements once experience with the new technique has been gained by the farmers (Barrett et al. 2014). An evaluation of rain-fed SRI production in West Bengal, India, calculated an 8% reduction in labour requirements (Sinha and Talati 2007). SRI study conducted in four districts of Tamil Nadu, India, found reduction in labour input by 17% and labour costs by 28% with the application of SRI methods (Barah 2009). However, the argument still remains inconclusive, with some of the recent studies pointing out that SRI is very labour intensive (Notlze et al. 2012). As pointed out by another recent study, the weak rate of adoption and stark rate of non-adoption in some regions are discussed to be because of SRI being labour intensive and requiring skill of farming (Palanisami et al. 2013). Surveys conducted at the household level and observations on fields in three rain-fed villages in the southern part of Cambodia showed a weak result of little or no difference between labour requirements by the adoption of SRI practices and conventional practices. The intensity of labour demand is needed when preparing land and nurseries, transplanting seedlings and harvesting. Because SRI is continuously evolving in the field, mechanised transplanting and mechanised weeders have also been designed in a number of countries; this may eventually reduce the labour requirements (Uphoff et al. 2011). Mechanised transplanting for SRI has been developed in countries like India, Costa Rica, Nepal, Iraq and probably some more countries by now (Uphoff et al. 2011). In countries like Pakistan, practically all SRI operations have been mechanised (Uphoff et al. 2011). These range from laser levelling of the fields to save water consumption by the plant, to constructing raised beds, to precision application of organic and inorganic fertilisers, to furrow irrigation (mostly done with siphons and gravity flow), to soil-aerating weeding and finally to harvesting.

However, there are some alternative views in terms of labour implications of SRI adoption. One of the latest studies by Gathorne-Hardy et al. (2016) showed that although SRI contributes to benefiting the environment and the economy, the socio-economic benefits that accrue to the farmer were at the expense of land-less labourers. The study showed that the labour demand for SRI is reduced, with the biggest decline in female employment. Additionally, SRI reduced casual labour pay per hectare by 50%. Therefore, policy support is essential for sharing private economic benefits of SRI with landless labourers.

The decline in female employment under SRI can be mainly due to the shift away from manual weeding operations to mechanical weeding that increased the demand for male labourers in weeding. There are differences in the manner in which weeds are treated under the traditional rice methods as well as SRI methods. In traditional method of rice cultivation, weeds are mostly removed manually, whereas under SRI weeds are removed mainly by using mechanical weeders. Under traditional methods of rice cultivation, weeds are seen as negative factor for crop growth, whereas under SRI weeds are considered as growth promoters when they are managed appropriately. Since the weeds are more in SRI due to intermittent wetting, it is important to manage the weeds regularly and this is done by mainly using the mechanical weeders.

Some studies have demonstrated the positive impact of weeder use on rice yields (Ramamoorthy 2004; Vijayakumar et al. 2006; Chadhar 2014). It is believed that mechanical weeding operations are better in facilitating aeration of the soil. This gives more room for better root and canopy growth. As the soil is kept moist but not inundated during vegetative growth phase, the soil is aerated and never become hypoxic. Early and frequent weeding is essential otherwise weed growth will become a problem. The study by Veeraputhran et al. (2014) for selected farmers' fields in Tamil Nadu state of India showed that adoption of SRI resulted in 11.06% higher yield, reduced weed management cost and better economic benefits which will pave the way for sustainable rice production and higher standard of living of the farming community in the study area. The objective of SRI is not only aimed at increased rice yield but also at promoting the higher productivity of land, labour, capital and water through efficient use of resources (Vijayakumar et al. 2006). Many countries like Indonesia, Madagascar and Bangladesh reported increase in rice yield with less water consumption under the SRI practices (Vijayakumar et al. 2006).

SRI's resistance to crop losses from pest and disease damage is often reported by farmers in China, India and Vietnam (Uphoff 2010). The SRI was first introduced in China in 1999. The experience shows that the SRI methods provide higher rice yield in comparison with current improved practices by more than 10% (Wu 2015). The experience shows that the higher yields are achieved with reduced requirements of water for irrigation and with reduced requirements for seeds. The reduction in water and seed usage is also leading to the reduction in farmers' costs of cultivation and therefore enhances their net income from rice production. Greater disease resistance, higher nitrogen use efficiency, enhanced photosynthetic rates and improved physiological traits, etc. are the other accompanied benefits of SRI cultivation. The experience of SRI in countries like Nigeria shows that SRI is introduced as a means to improve the rice productivity that could lead states like Cross River state to self-sufficiency in rice production (Bassey 2016). Therefore, the System of Rice Intensification (SRI) was demonstrated in all part of the state including Abi Local government Area.

The study done by Noltze et al. (2013) for Timor Leste shows that SRI has a positive impact on yield and household income. The study also shows that SRI is associated with higher family labour and management requirements, but with lower use of external inputs. Additionally, small farmers benefit more than the large farmers by adopting SRI. Nonetheless, in Timor Leste, SRI may not be beneficial when compared to conventional rice grown under favourable conditions. Another study done for Indonesia by Takahashi and Barrett (2013) shows substantial yield gains through adoption of SRI whereas there was no evidence for positive impact on household income. This was due to the fact that SRI induces a reallocation of family labour from non-farm to farm activities. Additionally, the study points out that the probability for SRI adoption increases with better water and labour availability and better social networks and decreases with the farmer's risk aversion attitude. The experience from Cambodia shows that the yield benefits of SRI remain inconclusive (Ches and Yamaji 2016). Some claim SRI increases rice yields at lower costs, while others state that the evidence is still inconclusive. Interestingly, the study shows that the key factor improving the degree of SRI adoption is farmer's zeal rather than the availability of labour (Ches and Yamaji 2016).

According to a study by Leesawatwong et al. (2005), SRI paddy usually gives a higher milling outturn, because when milled there is less chaff and less breaking of grains. Some studies highlight that the differences could vary even across plots; so a plot-specific study is important (Noltze et al. 2012). The study by Noltze et al. (2012) shows that the participation in training programmes and household labour availability increases the probability and intensity of adoption, as SRI is knowledge and labour intensive. While SRI was treated as labour intensive during the learning phase (Moser and Barrett 2003), in many countries, SRI practices are reducing farmers' requirements for labour as well as seed, water and production cost (Sinha and Talati 2007; Namara et al. 2008).

According to a study by Uphoff (2003), use of SRI methods can make agricultural systems more fruitful and favourable in the long run, by making irrigated rice production more sustainable and profitable. This is because SRI reduces the dependence on irrigation water by about half and reduces the requirements for capital and seeds. SRI methods are less dependent on external inputs like new-variety seeds, inorganic fertilisers or other agrochemicals and, therefore, have better prospects for sustainability. A study by Thakur et al. (2015) reveals that SRI methods showed significant improvements in rice phenotype and physiology. Integrating aquaculture and horticulture with SRI would further increase yield and water productivity. SRI methods make crop drought-tolerant and productive due to active root systems.

Several studies on SRI adoption point out various constraints in SRI adoption as well. Dobermann (2004), highlights a few points that are worth mentioning; (a) certain components of SRI that are perceived to be water-saving, like intermittent irrigation, require on the other hand, proper crop management and irrigation availability; (b) the adaptability of SRI may not be that easy owing to the agroecological conditions and environments in which SRI is evaluated. Perhaps, this could be a potential explanation behind the question of why farmers in several regions fail to adopt all components of SRI (Ly et al. 2012). Palanisami et al. (2013), nonetheless, supports the adoption of SRI by mentioning that studies conducted in the Indian sub-continent have shown that yield realisation under full adoption of SRI is significantly higher than partial adoption.

4.5 SRI in India

SRI in India has had a rather delayed start when we look at in comparison to when it began in Madagascar. Initial experiments with SRI in India were conducted in the year 2000 and that also only by organic farmers in Pondicherry (Prasad 2007). What triggered its use was the water-saving potential of SRI that attracted farmers from many southern states to this new method of farming (Basu and Leeuwis 2012). The need to improve water efficiency brought about a response by adopting some of the management practices of SRI while others were adapted to the local conditions (Senthilkumar et al. 2008). Gujja and Thiyagarajan (2009) note that in India SRI gradually at a slow pace became popular with farmers, and about one million hectares of area are under SRI cultivation, making it 2.42% of total area under rice cultivation in the country. Up to the present, field trials are being conducted in major rice-producing states of India like West Bengal, Punjab, Gujarat, Uttar Pradesh, Andhra Pradesh, Tamil Nadu, Karnataka, and there has been growing involvement of farmers, government institutions, research agencies and funding agencies to be in sync and work together for a large-scale adoption of SRI method in rice farming. Although data on actual levels of adoption and its impact are still not disseminated for India, the information that is accessible and public has attracted the attention of policymakers and planners in reviewing and promoting the SRI practices. Considerable efforts are being taken to promote SRI in different states. Out of 564 rice-growing districts in India, SRI is being practiced by the farmers in about 216 districts (ICRISAT 2008). Meanwhile, SRI is regarded as a key means of boosting national rice production under the government of India's National Food Security Mission (NFSM).

Palanisami et al. (2013) note that the available information does show that SRI has increased yield and has reduced consumption of water resource in selected locations. SRI study conducted in 13 major rice-producing states has shown that there are differences in output (yield), cost and gross margin between SRI and non-SRI fields. Table 4.2 gives a comparison between SRI and non-SRI fields among the sample farmers by Palanisami et al. (2013). The study indicates that, although there are inter-state variations, the SRI fields generally have higher yields as compared to non-SRI fields. For example, the average yield in SRI practices in all states is 8.5 quintals per hectare, which was 22% higher than in non-SRI fields. Madhya Pradesh, Gujarat and Odisha have significantly higher yields (in percentage terms it was 52, 54 and 33%, respectively).

Thus, SRI can be seen to provide increased yields in various agroecosystems with the use of fewer inputs such as water, seeds or fertiliser (Barah 2009; Zhao et al. 2009). A total of 2–3 kg of seeds per acre (as against the usual 30 kg) is used to raise the paddy nursery. Vasishth (2014) study using input–output model provides results and empirical evidence that the direct and indirect input requirements for rice cultivation wane when cultivation moves from conventional to SRI. SRI has shown to raise yields by changing farmers' agronomic practices towards more efficient use of natural resources, although sometimes this potential is neglected. A current report on Orissa (India) shows that the grain output was higher even though the cost of cultivation was 3.2% higher (SANDRP 2014).

Reddy et al. (2005) highlight that adoption rate of SRI has remained low despite several studies highlighting the high-yield and low-cost benefits of SRI. To understand this, one can mention that the major constraints in the adoption of SRI or modified SRI practice are the lack of skilled manpower for operations related to planting, weak control of water in the fields and unsuitable soils. Farmers risk

Table 4.2 Differences in yield, cost and gross margin between SRI and non-SRI fields

Region and State	Yield (q/ha)			Gross margin (Rs/ha)			Cost (Rs/q)		
	Non-SRI fields	Difference SRI-non-SRI		Non-sri fields	Difference SRI-non-SRI		Non-SRI fields	Difference SRI-non-SRI	
		Total	%		Total	%		Total	%
Southern region									
Andhra Pradesh	56.8	13	22.9	54,490	10,094	19	560	−146	−26
Karnataka	56.7	14.1	24.8	56,277	12,885	23	370	56	15
Kerala	47.1	6.4	13.6	51,613	7044	14	857	−244	−28
Tamil Nadu	45.5	6.8	14.9	41,879	5786	14	675	−223	−33
Average	51	9.2	18	49,552	8290	17	617	−168	−27
Western region									
Gujarat	18.7	10	53.6	17,274	8973	52	757	−234	−31
Maharashtra	27.9	7.5	26.9	26,904	4266	16	527	−253	−48
Rajasthan	20.9	2.7	12.9	41,145	5327	13	2068	−201	−10
Average	25.6	7.3	28.5	27,597	6585	24	715	−220	−31
Eastern region									
Chhattisgarh	48.7	11.9	24.5	53,451	1257	2	581	−167	−29
Odisha	36.2	12	33.1	33,929	12,111	36	669	−151	−23
Uttar Pradesh	54.5	8.5	15.5	53,655	8334	16	655	−41	−6
West Bengal	36	5.9	16.5	32,885	5400	16	507	−14	−3
Average	40.9	9	22	38,446	7474	19	585	−71	−12
Central region									
Madhya Pradesh	19.3	10	51.9	12,530	11,184	89	430	−56	−13
North Eastern region									
Assam	34.1	4.1	12	32,188	3504	11	674	−380	−56
All India	37.9	8.5	22.4	37,845	6971	18	621	−178	−29

Source Palanisami et al. (2013)

aversion also plays a role behind the reason for poor adoption (Johnson and Vijayaraghavan 2011).

Devi and Ponnarasi (2009) study (via a logistic regression model) showed that the low adoption behaviour in Cuddalore district of Tamil Nadu was because of the lack of skilled labourers, awareness, training on new technology, farm income, experience, etc. A macro-level study covering 13 major rice-producing states using multinomial logit model showed that the factors such as soil type and source of irrigation were important in influencing the level of adoption in all the four regions (Southern, West, East, Central, North-East) (Palanisami et al. 2013). Study conducted in the Bardhaman district of West Bengal (via a logit regression model) highlights that the education level and distance from the canal were contributing factors in the constraints to adopt SRI (Haldar et al. 2012). As per the study done by Islam et al. (2014) for Meghalaya, the major contributors to the non-adoption of SRI were related to involvement of more efforts, faith towards traditional practices, ignorance and lack of knowledge of scientific water management.

4.6 Farmers' Experience of SRI in Karnataka

Before undertaking a final field survey, a pilot survey was conducted in 3 villages, namely Saraguru, Banugondi and Kabaligere, and 11 households cultivating paddy as their primary crop were visited. Based on the pilot study and interaction with the Community Area Development Authority (CADA) directors and farmers of Banugondi and Saraguru villages of Arkalgud taluk in Karnataka, some important insights were gained. An increase in yield was observed in case of SRI method as compared to the traditional methods of rice cultivation. This difference in yield can be attributed to the difference in land pattern and type of seeds. Wetland provided higher yield than dryland, and high-yield variety of seeds provided a higher yield than the hybrid seeds. It was observed that cost of labour was higher under SRI than the traditional method of cultivation. Farmers who adopted SRI method of cultivation faced labour problems such as lack of skilled and trained labourers. The supervision of labour required in SRI method is tedious and time-consuming.

Hassan District

The SRI was introduced in all talukas of Hassan district in 2005–2006. In Hassan, SRI was also incorporated under the government's NFSM. SRI is promoted by the agricultural department and in order to encourage farmers in adopting SRI, the government department provides various input subsidies such as seeds, fertiliser and farm equipment to them. The department made use of the progressive farmers to take initiative in SRI adoption and also to spread awareness about the method. Pioneering farmers who adopted SRI observed substantial yield differences between conventional and SRI practices, and this led to the wider adoption of SRI in the district. The farmers also observed differences in water requirements between the two practices: SRI required less water as compared to the conventional practice. In their view, SRI method was beneficial but the major bottleneck in its adoption was the lack of availability of labourers; however, this was true in the case of conventional rice farming practices as well. The farmers opined that the young farmers were having greater access to off-farm employment, and this resulted in lower supply of young labourers for rice farming. Besides, labourers also prefer to work in coffee plantations as they get a higher wage rate in the coffee estates. Farmers also find rice cultivation as a risky enterprise due to the uncertainty in market prices. Also, partial adoption of SRI was quite common, and most farmers did not follow the practice of one seedling at a wider space. Farmers had the tendency to plant more than one seedling and reduce the space between the seedlings as a measure to shield themselves from crop loss. Also, in the heavy rainfall areas, the alternate watering and drying was not followed by many of the farmers. However, there were few who overcame this limitation through bunding their paddy plot, thus restraining the heavy flow of water through their field. Also, farmers opined that the market prices were the same for both SRI rice variety and conventional rice variety.

Chikmagalur District

In sharp contrast to Hassan, in Chikmagalur, NGOs were more active in promoting the SRI. For example, Shri kshetra Dharmasthala Rural Development (SKDRDP) was active in promoting SRI in Karnataka in general and in districts like Chikmagalur, where the SRI is not incorporated under NFSM, in particular.

Several farmers in Chikmagalur who had adopted SRI earlier discontinued the practice. They replaced their land under rice for coffee cultivation. Coffee cultivation seemed a more attractive option for farmers in Chikmagalur. Besides rice and coffee, pepper and areca nut were also popular in the district. However, the market for the sale of produce was usually the local market.

Several farmers cultivated rice mainly for self-consumption rather than for sale in the market. Farmers also opined that it was difficult to find labourers for SRI as many labourers find SRI to be a tedious job. Not only that, the labourers demanded relatively higher wages for SRI as compared to traditional rice cultivation. Another interesting observation from the field was that labourers were more attracted to coffee plantation sector as they were getting higher wages there.

Farmers in Chikmagalur district had come up with some strategies to overcome the limitations of labour. Through mutual understanding among farmers, they formed groups and worked alternatively on farms of one another. By doing this, they tried to reduce the dependence on labourers. This was observed in the hilly regions of Chikmagalur.

As far as some practices of SRI are concerned, farmers from the heavy rainfall and hilly region (Kalasa, Mudigere) were not able to follow the practice of alternate drying and wetting of plot. Also, they did not follow the one seedling per hill practice. However, the spacing between and across the seedling was followed by most of the SRI farmers in Chikmagalur district.

4.7 Farmers Experience of SRI in Madhya Pradesh

Shahdol District

The SRI practice was promoted in the district since 2012. Initially, it was entirely promoted under NFSM; however, from the past one year, it is being promoted under Rashtriya Krishi Vikas Yojana (RKVY) along with NFSM. Three blocks in Shahdol district, namely Gohparu, Sohagpur and Jaisinghnagar were covered, consisting of a total of 64 SRI and non-SRI farmers. The primary crop for the Kharif season is paddy along with other crops such as maize, tur and til. The farmers with conducive land structure and irrigation facility cultivated vegetables such as potato, tomato and ladyfinger. These crops were cultivated mainly for consumption purpose, and the surplus was sold in the nearest market. However, it was observed that the entire production of paddy was needed for consumption and barely any quantity was sold in the market. It was also observed that availability of water at the time of germination and growth of seedling has immense importance. Majority of the farmers were totally dependent on rainfall and did not have any other irrigation facility. Inadequate rain delayed the transplantation process which resulted in decrease in yield. Lack of irrigation facility restricted the farmers from adopting SRI. Labour issues in the district were minimal as farmers were dependent entirely on family labour. Transplantation work was done by female labourers who found SRI plantation practice less strenuous than the conventional practice. Lower input requirement, particularly with respect to seed requirement, and minimal usage of chemical fertilisers were some of the benefits of SRI plantation. Biogas plants provided by the Department of Agriculture were functional in a few parts of Sohagpur block. In the case of non-SRI farmers, majority of them cultivated paddy through broadcasting method under which local or traditional variety of paddy was cultivated. It was observed that both SRI and non-SRI farmers were dependent on PDS. However, irregular distribution of pre-decided benefits and non-accessibility of fair price shops were the primary problems.

Sidhi District

Pradan has been promoting SRI practice within the district for the past 4–5 years. They provide training on SRI cultivation in Sidhi and Sihawal blocks. Training is provided by the community service provider (CSP) at individual village level. Sidhi and Sihawal blocks consisting of a total of 64 SRI and non-SRI farmers were covered. The cropping pattern in Sidhi was similar to Shahdol. Paddy is the primary Kharif crop with other subsidiary crops such as maize. Some of the benefits reaped by the framers include an increase in total number of tillers, less weeding requirement, less seed requirement under SRI method and use of bio fertilisers. Despite these benefits, the major constraints faced by the farmers include lack of irrigation facility and delay in transplantation due to irregular rainfall, leading to a decline in yield. Due to lack of such equipments as marker, SRI farmers were not able to maintain proper spacing between the plantings. Also, due to lack of equipments such as cono-weeder and uncontrolled water supply, farmers were unable to practice alternate drying and weeding.

4.8 Farmers Experience of SRI in Orissa

Keonjhar District
The SRI practice in Keonjhar district was initiated in 2008. SRI was adopted in 13 blocks of Keonjhar district. The study covered Sadar, Patna and Harichandanpur blocks of Keonjhar district. The study interviewed 50 SRI and non-SRI farmers in the district. In order to promote SRI adoption, the Agriculture department provides subsidy on farm equipments such as cono-weeder and marker. This practice was promoted at the block level with the field officers at individual blocks who provided training to the farmers. Paddy is the primary crop in Keonjhar district cultivated to meet subsistence needs of farmers in the district. The surplus, if left any, is sold in the market. In most areas, local variety of seeds are cultivated by the farmers. Livestock rearing and poultry are the leading sources of income in the district. In addition to farming, everyone worked as NREGA labourers and availed the benefits of BPL and AD cards. The yield and cost difference between SRI and conventional method of cultivation was observed by the farmers. Scarcity of farm labour and inadequate water availability were the biggest constraints faced by the farmers for adoption of SRI and conventional practice. Areas in the district which were entirely dependent on rainfall were not able to adopt alternate drying and weeding practice, resulting in crop losses.

Mayurbhanj District
Paddy is the primary crop in the district cultivated for self-consumption. Besides paddy, farmers also cultivate vegetables such as brinjal, tomato and potato. The study covered 2 blocks from Mayurbhanj district. The study interviewed 40 SRI and non-SRI farmers in the district. It was revealed that farmers had been practicing SRI for the past 5–6 years. The yield and cost difference was observed by the farmers, and accordingly, farmers have allotted part of their landholding to SRI. Farmers in the district cultivate hybrid seeds along with local variety of seeds. Inadequate labour, irregular rainfall and lack of appropriate landholdings for SRI cultivation led the farmers to discontinue SRI practice. Farmers in the district depend heavily on cattle rather than on tractors and tillers. In addition to farming, livestock rearing and poultry are the leading sources of income in the district. In addition to farming, everyone worked as a NREGA labourer and availed the benefits of BPL and AD cards.

4.9 Gender Implications of SRI

An estimated 500 million women worldwide are involved in rice cultivation. It is a labour-intensive activity and is usually performed in unsanitary conditions. The decline in profitability of farming, particularly rice farming, has resulted in men migrating to towns and cities in search of jobs. This has resulted in women taking up the responsibility of farming. Women benefit from using SRI practices in many ways, explaining why SRI is gaining popularity around the world.

A study of the gender effects of SRI adoption in Cambodia, commissioned by Oxfam America, found that female adopters of SRI reported that purchasing fewer seeds and using fewer fertilisers resulted in savings. Also, the reduced use of inputs has lightened the burdens of farming which give family members more time to do other things. Interviews revealed that men helped women less in transplanting and uprooting rice seedlings since these were fewer and tasks became lighter with SRI. Selecting seeds and transplanting seedlings are tasks generally assigned to women, and SRI made these tasks lighter for women and has provided them with more time for domestic work, paid work on other farms and backyard livelihoods (Resurreccion et al. 2008). Traditionally, weeding was done by women by hand. However, in SRI a mechanical hand weeder is used. With the use of mechanical weeders, women's labour time and physical discomfort from weeding have reduced. However, since cultural norms expect men to do mechanical work, in some parts of India men have taken over the task of SRI weeding. A study in Tamil Nadu found that men's labour in rice cultivation has increased by 60% while women's workload in rice cultivation has reduced by 25%.

4.10 Key Actors and the Role of Institutions in the Spread of SRI

Unlike other innovations, SRI is a farmer centric innovation and its success depends largely on farmers' motivation and skills. A wide array of institutions—national as well local—plays a role in the wider dissemination and application of SRI and its principles. They include government agencies, non-governmental organisation, universities and research institutions, as well as private sector organisations (e.g. Syngenta in Bangladesh, Nippon Koei in Indonesia) (Uphoff and Kassam 2009).

The study by Kumar and Saha (2013) in selected states of India—Bihar, Chhattisgarh, Jharkhand and Orissa—attempts to evaluate the initiative taken by the state governments, Civil Society Organisations (CSOs), research institutions and other stakeholders in promoting the adoption of SRI. The study shows that the CSOs have contributed in a big way to influence several state governments to create a conducive environment for the adoption of SRI on a large scale. For example, in Bihar, the SRI programme began mainly because of the groundwork of CSOs. Later, the agriculture departments of state governments, the Jeevika programme and CSOs took it to a larger scale. The study points out that in other states also the SRI programme was started by CSOs, and the government's response has been minimal. Additionally, in states such as Jharkhand and Chhattisgarh, government support for the dissemination and adoption of SRI has been passive. The study by Kumar and Saha (2013) further notes that in Bihar, the synergy between the non-governmental organisations—PRADAN—and Bihar Rural Livelihoods Promotion Society (BRLPS) made a great impact, whereas in Jharkhand, the synergy between National

Bank for Agriculture and Rural Development (NABARD) and CSOs yielded positive results. In Chhattisgarh, paddy procurement, revamped by computerisation, has resulted in very good market selling price (MSP) for farmers (Kumar and Saha 2013).

In the three states that we analysed, we observed that NGOs were more active in those districts where the SRI is not incorporated under the NFSM whereas, in those districts where SRI is incorporated within the NFSM, the agricultural departments were more active in promoting SRI. Shri Kshetra Dharmasthala Rural Development Project (SKDRDP), an NGO working for rural development in Karnataka, initiated steps to popularise SRI method of paddy cultivation across Karnataka. In Dakshina Kannada district, 937 farmer families have implemented the SRI method in 939 acres of land. In Orissa, the government has taken up line-sowing on a large scale (Kumar and Saha 2013). Also in order to promote SRI, a State Plan scheme has been implemented in the state from 2009 to 2010 with the objective of enhancing rice production by increasing water use efficiency and economising use of quality seeds and other inputs. In 2014–2015, SRI was taken up in 18,245 ha with an expenditure of Rs. 912.25 lakh, and the government also aimed to promote SRI in 20,000 ha with an estimated cost of Rs. 1000.00 lakh during 2015–2016 (Activity Report 2014–15). The experience of SRI in Madhya Pradesh shows that the SRI system of rice cultivation has become widespread where controlled irrigation is easily available; the area under rice cultivation has gone up to 1.8 million hectares in the year 2014.

A study by Prasad et al. (2008) on SRI in Orissa shows that in the state, SRI was first experimented by a few NGOs in 2005–2006 and subsequent to this the first initiative by the government took place in the year 2007. This was during the period when demonstrations on this new practice were incorporated under government schemes like Rashtriya Krishi Vikash Yojana (RKVY) and Agriculture Technology Management Agency (ATMA). Prasad et al. (2008) highlights that a learning Alliance initiated by the Xavier Institute of Management, Bhubaneswar, was the precursor to such widescale interventions. Following this, the demonstrations and field experiments have been incorporated into all other schemes for rice development, leading to widespread adoption of SRI in the state.

Several trainings were also initiated at the field level to enhance the knowledge among different stakeholders. In addition to this, an impact evaluation at the farmer field level was undertaken by the state Agriculture Department. Studies conducted by CRRI (Central Rice Research Institute), Cuttack, and Agronomy Department of OUAT (Orissa University of Agriculture and Research) also proved the benefits of introducing SRI practices in the state (Prasad et al. 2008). In the year 2007–2008, around 80 officers of State Agriculture Department were trained by the CRRI, OUAT, and IMAGE (the State Agriculture Extension Management and Training Institute).

The concept of SRI village is also an important feature of the history of SRI development in Orissa. The Agriculture department as an innovative measure adopted one revenue village in each of the 30 districts to convert the paddy plots in the selected villages into SRI practice (Prasad et al. 2008). These SRI villages will become the focal point for quick dissemination of this new practice (Table 4.3).

Table 4.3 SRI organisations in Orissa

S. No.	Organisation	District	No. of blocks	No. of villages	No. of farmers practicing SRI	Estimated SRI coverage area (ha)
1	Pragati	Koraput	5	86	266	56.76
2	United Artists' Association	Ganjam	3	42	711	*NA*
3	CARR	Cuttack	1	*NA*	20	2.91
4	OPDSC	Rayagada	5	10		50
5	BOJBP	Nayagarh	5	50	376	88
6	SACAL	Gajapati	1	26	59	11
7	Sahabhagi Vikas Abhiyan	Kalahandi, Bolangir Nuapada	13	126	1100	254.25
8	MASS	Sambalpur & Bargarh	3	19	8	5.75
9	PRADAN	Mayurbhanj	3	*NA*	1688	210
10	Sambhav	Nayagarh	*NA*	*NA*	*NA*	
11	Karrtabya	Kalahandi	3	23	108	42.29
12	WORD	Keonjhar	1	6	5	0.37
13	VICALP	Ganjam, Gajapati	2	NA	10	5
14	RUPA	Dhenkanal	1	7	4	1.25
15	Vasundhara	Nayagarh	1	9	29	2.02
16	Madhyam Foundation	Khurda	1	*NA*	82	27.2
17	NIRMAN	Nayagarh	4	*NA*	28	4.16
18	Centre for Community Development	Gajapati	1	5	34	6
19	SPARD	Cuttack	1	2	8	1.85
20	SOLAR	Puri	1	3	*NA*	*NA*
	Total		46	414	4536	768.81

Note NA data not available
Source Prasad et al. (2008)

4.11 Conclusion

The literature reviewed in general as well in the Indian context shows mixed results for SRI, although majority of the studies show SRI as beneficial to rice farmers. The economic, social and environmental benefits of SRI vary quite considerably across the regions. Since the SRI does not have a unique definition of practices, the method faces some methodological challenges as well as problems in the evaluation of performance.

The farmers' experience of adoption of SRI based on our field-level observations pointed out that most farmers agree with the higher yield benefits of SRI as compared to traditional rice cultivation. The farmers also opined that the input requirements are relatively less. However, availability of labourers for both SRI and rice cultivation was a major cause for concern. In some districts such as Shahdol, the labour issues were minimal due to the fact the farming was heavily dependent upon the family labourers, whereas the labour issues were relatively higher in Karnataka. Another interesting observation from farmers' experience was the difficulty in adopting water management—alternate wetting and drying—due to heavy rainfall. So those farmers who were dependant on rainfall for rice cultivation were not able to adopt the water management practices even when they adopted other practices of SRI. Therefore, partial adoption was common. In some districts like Shahdol, inadequate rainfall or delay in rainfall was also a major cause for concern. Another major lacuna in adoption was the lack of equipment such as cono-weeder, thus discouraging the farmers from adopting cono-weeders for weeding which is also an important component of SRI. The government's support and subsidy for equipment such as cono-weeder may not be sufficient.

The next chapter after discussing the salient features of NFSM and socio-economic profile of the sample households will analyse the adoption of SRI and the impact of SRI adoption on household income and yield by employing diverse approaches and methods.

Chapter 5
National Food Security Mission and SRI

5.1 Introduction

The National Food Security Mission (NFSM) was launched by the government of India during 2007–2008 at the beginning of the 11th Five-Year Plan. The major objective of the programme was to address the issue of food security by devising programmes targeted to escalate production of rice, wheat and pulses by 10, 8 and 2 million tonnes, respectively, by the end of the 11th Five-Year Plan. The NFSM was initially implemented in 482 districts of 19 states comprising of 144 districts under rice in 16 states, 142 districts under wheat in 9 states and 468 districts under pulses in 16 states.

Under NFSM-Rice, six north-eastern states—Arunachal Pradesh, Manipur, Meghalaya, Mizoram, Nagaland and Sikkim—have been added since 2012–2013. Similarly, two hilly states, viz. Himachal Pradesh and Uttarakhand under NFSM-Rice and -Wheat and J&K under NFSM-Wheat, were also added in the same year. As a result the number of states that implemented NFSM increased to 27 in the year 2012–2013. NFSM-Rice is being implemented in 210 districts of 24 states, NFSM-Wheat in 166 districts of 12 states and NFSM-Pulses in 468 districts of 16 states. Figure 5.1 shows the states covered under NFSM-rice programme.

The NFSM had a two pronged strategy. First strategy was to expand the area, and the second strategy was to enhance the productivity by bridging the gap between the actual and potential yield. However, the area expansion was confined to mainly wheat and pulses, and in the case of rice, the focus was mainly in enhancing the productivity without much of area expansion (Manjunatha and Kumar 2015). In order to augment productivity, the NFSM had adopted several measures including (1) quality seed production; (2) emphasising integrated nutrient management and integrated pest management; (3) promotion of new technologies; (4) restoring soil fertility; and (5) improved farm implements. As a result, a total

P. Varma, *Rice Productivity and Food Security in India*,
DOI 10.1007/978-981-10-3692-7_5

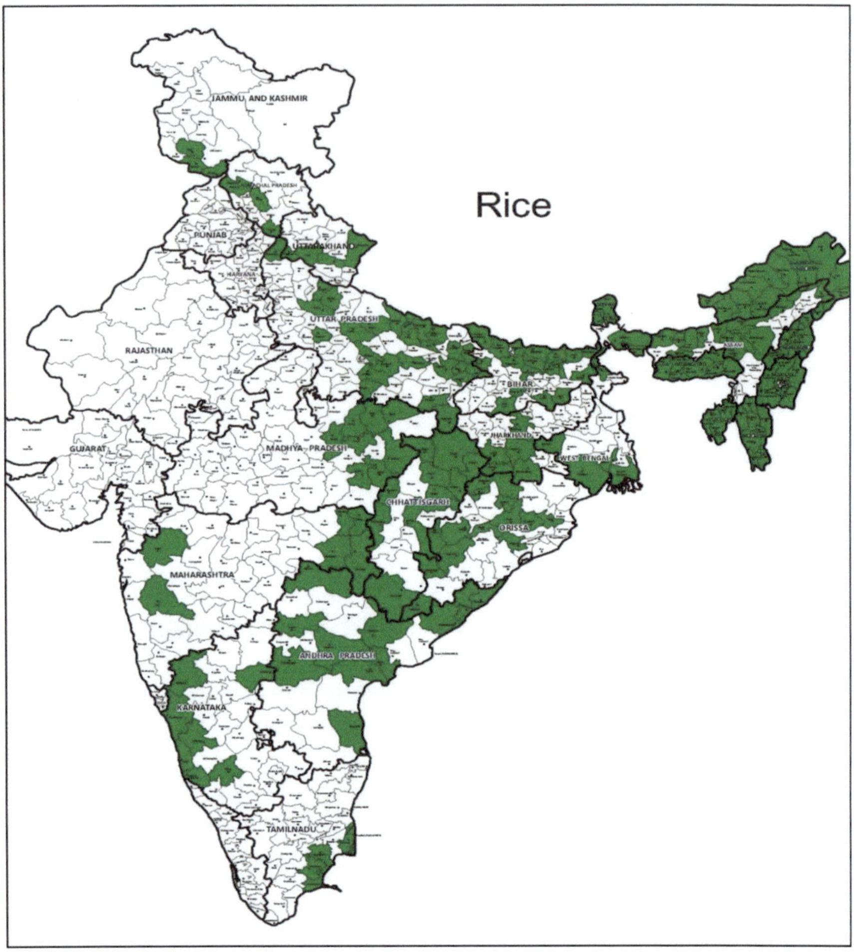

Fig. 5.1 States covered by NFSM for rice. *Source* http://nfsm.gov.in/Riceindia.pdf

amount of Rs. 4500 crores have been spent under NFSM during the 11th five year plan (Manjunatha and Kumar 2015).

Rashtriya Krishi Vikas Yojana (RKVY) programme was also introduced during the same period when NFSM was introduced. Apart from NFSM, several states and centrally sponsored programmes were also added impetus to the food security promotion programmes. As a joint results of all these, the wheat production during the end of the 11th Five-Year Plan increased by 19.1 million tonnes, paddy production increased by 12.1 million tonnes and pulses production by 3 million tonnes as compared to the year 2006–2007.

5.2 NFSM-Rice Interventions

In order to improve rice productivity, the following interventions have been put into operation.

> Demonstration of Improved Package of Practices, Demonstration on System of Rice Intensification, Demonstration on Hybrid Rice Technologies, Support for Promotion of Hybrid Rice Seed and HYV Seed, Incentive for Micronutrients, Incentive for Liming in Acidic Soils, Assistance for Plant Protection Chemicals and Bio-agents, Incentive for Cono-weeders and other Implements, Incentive for Zero till Seed Drills, Incentive on Multi Crop Planters, Incentive on Seed Drills, Incentive on Power Weeders, Incentive on Rotavators, Incentive for Pump Sets, Farmers' Field School, Knap Sack Sprayers, Laser Levellers, Miscellaneous and Local Initiatives (Agricultural Finance Corporation Limited 2012).

5.3 NFSM-SRI Interventions

There were field demonstrations of SRI in 14 states. For example, out of the targeted field demonstrations of 68,043, the states were undertaken around 68,856 field demonstrations of SRI. Seed distribution, application of micronutrients/lime, plant protection chemicals, distribution of farm implements and extension programmes were the part of interventions under NFSM-Rice. Out of the total interventions, 36% of the share was devoted to the distribution of seed, which had the maximum share, and this was followed by the application of micronutrients/lime with 31% share, plant protection chemicals with 17%, distribution of farm implements with 15% and extension programmes having 2% share. Seed and micronutrient subcomponents were the 2 main activities forming about two-thirds of the total programme (Agricultural Finance Corporation Limited 2012).

Several extension services were conducted under the programmes of Farmers' Field Schools (FFS) and frontline demonstration programmes. Frontline demonstration programmes are organised to demonstrate yield gain due to improved package of practices, hybrid rice production technology and yield gain under SRI. The targeted extension service programmes by the states were around 2.50 lakh during 2007–2008 and 2010–2011. Against this target, 2.38 lakh FFSs and frontline demonstration programmes were held in all the states. The overall achievement at 95% of the target is fairly satisfactory (Agricultural Finance Corporation Limited 2012). A summary of NFSM interventions and pattern of assistance for rice are given in Table 7.8.

5.4 Impact of NFSM on Rice

The results from the NFSM show that the NFSM districts recorded higher rice yield gains as compared to non-NFSM districts. For example, the yield gain in rice was about five times more than the yield obtained in the non-NFSM districts during 2007–2008 and 2008–2009. For example, out of the districts initially covered under NFSM-rice programme, 83, 82 and 70 districts recorded higher yields during 2007–2008, 2008–2009 and 2009–2010, respectively. The reduction in number of districts recording higher yield during 2009–2010 can be attributed to severe to moderate drought conditions during the year in many states including Bihar, Jharkhand, Chhattisgarh, Madhya Pradesh, parts of Uttar Pradesh, Andhra Pradesh, Orissa and parts of Karnataka (Agricultural Finance Corporation Limited 2012).

However, as far as the area under rice is concerned, the non-NFSM districts performed better area expansion during 2007–2008 and 2008–2009. But during the drought conditions in 2009–2010, the NFSM districts performed better than the non-NFSM districts with relatively less adverse effect in the overall decrease in the area put under rice cultivation. The decrease in area in NFSM districts was 5.71% against 6.17% in non-NFSM districts. The package of improved agricultural practices recorded a higher growth rate of 3.99 and 7.40% during 2007–2008 and 2008–2009 in NFSM districts as against 2.91 and 5.48% in non-NFSM districts. The effect of drought on decline in production was less severe in NFSM districts (5.44%) compared with 8.79% decline in production in the non-NFSM districts.

The National Food Security Mission (NFSM) is in operation in 27 states of the country including Karnataka, Madhya Pradesh and Orissa. The details of the states covered under NFSM in these three states are given in Table 5.7.

5.5 NFSM in Karnataka

National Food Security Mission-Rice (NFSM-Rice) and National Food Security Mission-Pulses (NFSM-Pulses) are being implemented in the state. The area, production and yield of rice in both NFSM and non-NFSM districts are given in Tables 5.1 and 5.2. In Karnataka, Hassan is a NFSM district, whereas Chikmagalur is a non-NFSM district. The data shows that the average rice yield over the six-year period was marginally higher in NFSM district (Hassan) than the non-NFSM district (Chikmagalur). The average rice yield in Hassan during this period was 2498 kg/ha, whereas in Chikmagalur, the average rice yield was 2479 kg/ha.

Area under rice cultivation was slightly higher in Hassan as compared to Chikmagalur. When Hassan experienced a marginal improvement in area under rice between 2011–2012 and 2012–2013, Chikmaglaur experienced a decline in area under rice cultivation, and as a result, the production is also gone down in Chikmagalur (see Tables 5.1 and 5.2).

Table 5.8 (continued)

3.7	Multicrop planter	Rs. 15,000/- per machine or 50% cost whichever is less
3.8	Zero till multicrop planter	Rs. 15,000/-per machine or 50% cost whichever is less
3.9	Power weeder	Rs. 15,000/- per machine or 50% cost whichever is less
3.10	Water-carrying pipes	Rs. 25/- per metre or 50% of the cost whichever is less with maximum limit of 600 m and costing Rs. 15,000/-
3.11	Rotavator/turbo seeder	Rs. 35,000/- per machine or 50% of cost whichever is less
3.12	Paddy thresher/multicrop thresher	Rs. 40,000/- per machine or 50% of the cost whichever is less
3.13	Laser land leveler	Rs. 1.50 lakh per machine to a group of 10 farmers
3.14	Self-propelled paddy transplanter	Rs. 75,000/- per machine or 50% of the cost whichever is less
4. Plant protection		
4.1	Plant protection chemicals and bio-pesticides/IPM	Rs. 500/- per ha or 50% of the cost whichever is less
4.2	Weedicides	Rs. 500/- per ha or 50% of the cost whichever is less
5. Micronutrients and Soil ameliorants		
5.1	Micronutrients	50% of the cost limited to Rs. 500 per ha
5.2	Lime/liming materials	50% of the cost material limited to Rs. 1000 per ha
6.	Local initiatives	Funding will be on project basis, up to 5% of the total allocation to the state
7. Project management teams at state and district level		

Source http://nfsm.gov.in/rice.aspx

production declined during 2007–2008 to 2010–2011 and after that showed an improvement. However, the production went down from 564 thousand tonnes in 2007–2008 to 519 thousand tonnes in 2012–2013. The yield of rice declined sharply between 2007–2008 and 2010–2011 and after that showed an improvement. This improvement can be attributed to an improvement in production that took place since 2010–2011 (see Table 5.6). The yield for rice was 1649 kg/ha in 2007–2008, and it improved to 1834 kg/ha in 2012–2013 (see Tables 5.6, 5.7 and 5.8).

Chapter 6
An Overview of SRI Adoption and Socio-economic Profile of the Households

6.1 Introduction

This chapter provides an overview of the adoption of SRI across different states and discusses the socio-economic profile of the sample households. Considering the heterogeneous nature of the country and the study region, there were considerable differences across the states in terms of the intensity and the depth of adoption of SRI. The sections below undertake a discussion of SRI adoption across states, the socio-economic profile of the total sample size, the detailed socio-economic profile at the district level, etc. A detailed socio-economic profile at the district level is undertaken to understand the disparities in terms of various social, economic and institutional factors across different states. This chapter will also provide a descriptive analysis of yield, cost and income differences between adopters and non-adopters of SRI.

6.2 An Overview of SRI Adoption in the Selected Study Areas

The intensity of SRI adoption is the highest in Shahdol district of Madhya Pradesh which shows that 50.76% share of total land under rice has been allocated to SRI. This is followed by Hassan district of Karnataka which devoted around 40% of total land under rice for SRI. The huge disparities in the level of adoption were observed in Madhya Pradesh: Shahdol had 50.76% while Sidhi had only 36.8%. The lowest adoption was noted in Mayurbhanj district of Orissa. In Mayurbhanj, the acres devoted for SRI in total land was only 30%. Therefore, the percentage share of area under SRI in total land under rice is the highest in the state of Madhya Pradesh and the lowest in the state of Orissa. The disparities in the level of adoption were the lowest in Karnataka state (see Fig. 6.1).

P. Varma, *Rice Productivity and Food Security in India*,
DOI 10.1007/978-981-10-3692-7_6

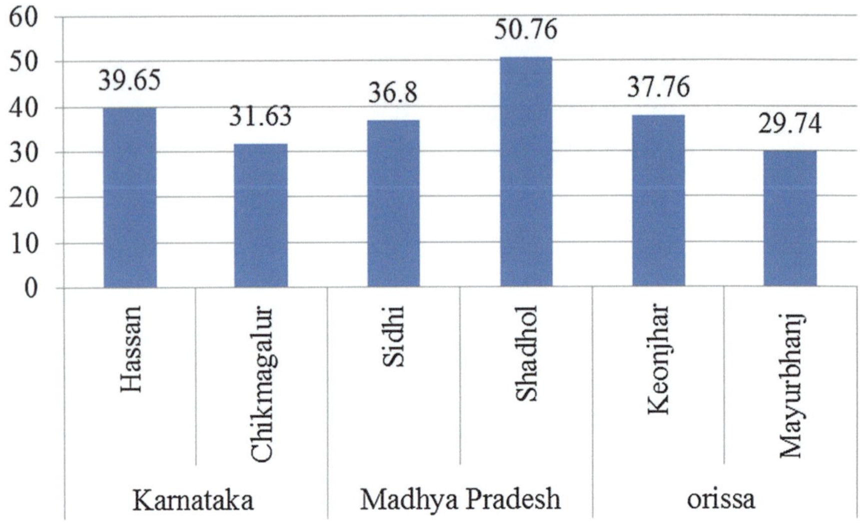

Fig. 6.1 Intensity of SRI adoption (% share of area under SRI in total rice). *Source* Survey data

The SRI emerged as a set of 6 practices. They are young seedling, shallow planting, single seedling at a wider spacing, use of organics, use of cono-weeder and alternate wetting and drying. In this study, we have decided to split the third practice into two—single seedling and wider spacing. This is due to the fact that we observed that many farmers, although allowing wider spacing, were planting more than one seedling.

The adoption was the highest for shallow planting followed by young seedling and then wider spacing. Around 180 farmers had followed shallow planting and 156 farmers followed young seedling and 147 farmers followed wider spacing. The adoption was the lowest for use of organics, with only 66 farmers adopting the use of organics. This was followed by alternate wetting and drying, with only 109 farmers following this practice (see Fig. 6.2).

Figure 6.3 shows the percentage of farmers who adopted only one practice, two practices and so on. The figure shows that the percentage of farmers who adopted 5 SRI practices was the highest, followed by 6 practices and 4 practices. Using the information from Fig. 6.2, we can assume that most farmers had adopted 4–5 practices, and also farmers might have adopted the plant management practices (young seedling, shallow planting, single seedling, wider spacing, etc.) more than soil management and water management practices (Fig. 6.4).

Further analysis of data shows that all the farmers in Hassan district of Karnataka had adopted the practices of wider spacing and organics. Similarly, in Chikmagalur district of Orissa, all the farmers had adopted young seedlings and cono-weeder. All the SRI farmers in Shahdol had adopted shallow planting (see Table 6.1).

Also, planting young seedling was the highest in Chikmagalur followed by Mayurbhanj and Hassan. Shallow planting was the highest in Shahdol followed by

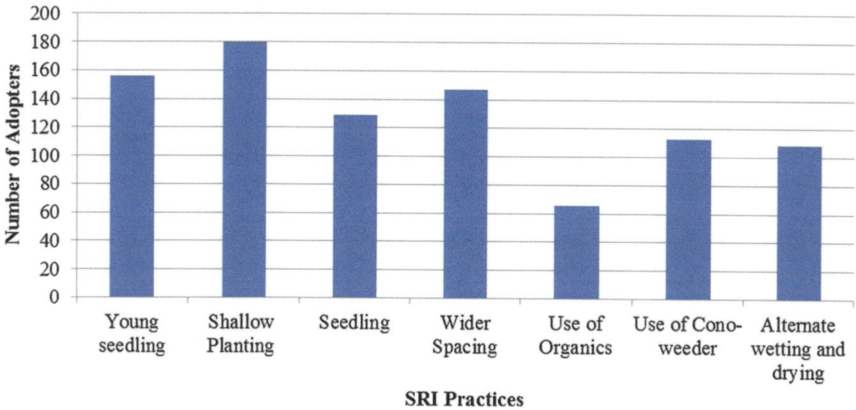

Fig. 6.2 Number of adopters for each SRI practice. *Source* Survey data

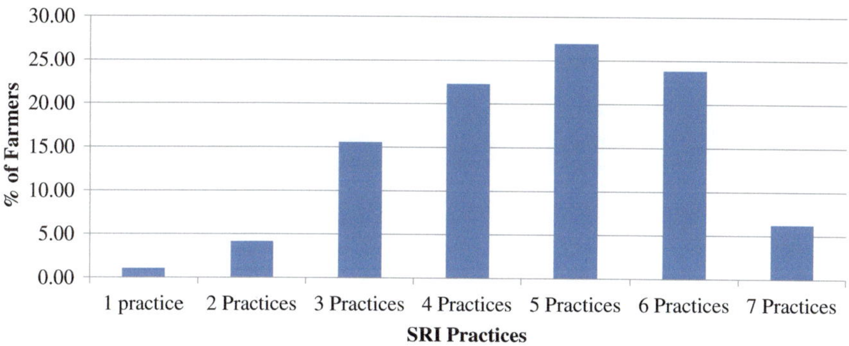

Fig. 6.3 Percentage of depth of SRI practices adoption. *Source* Survey data

Keonjhar, Sidhi and Mayurbhanj. Single-seedling practices were the highest in Keonjhar, followed by Shahdol and Mayurbhanj. Practices of wider spacing were the highest in Hassan, followed by Chikmagalur. Use of organics was the highest in Hassan. The use of organics was relatively less in other districts. Use of cono-weeder was the highest in Chikmagalur, followed by Hassan. Use of cono-weeder was the lowest in Sidhi. Alternate wetting and drying was the highest in Shahdol and lowest in Sidhi (see Table 6.1).

Analysis of adoption level across districts showed that full adopters of SRI were the highest in Hassan and Shahdol and the lowest in Mayurbhanj (see Fig. 6.2). We define adopters of all 6 practices as full adopters, adopters of more than 3 but less than 6 practices as partial adopters, and adopters of 3 or less practices as low adopters. Partial adopters of SRI were the highest in Keonjhar followed by Mayurbhanj. Partial adopters of SRI were the lowest in Hassan. Low adopters of SRI were the highest in Sidhi followed by Mayurbhanj. Low adopters were very

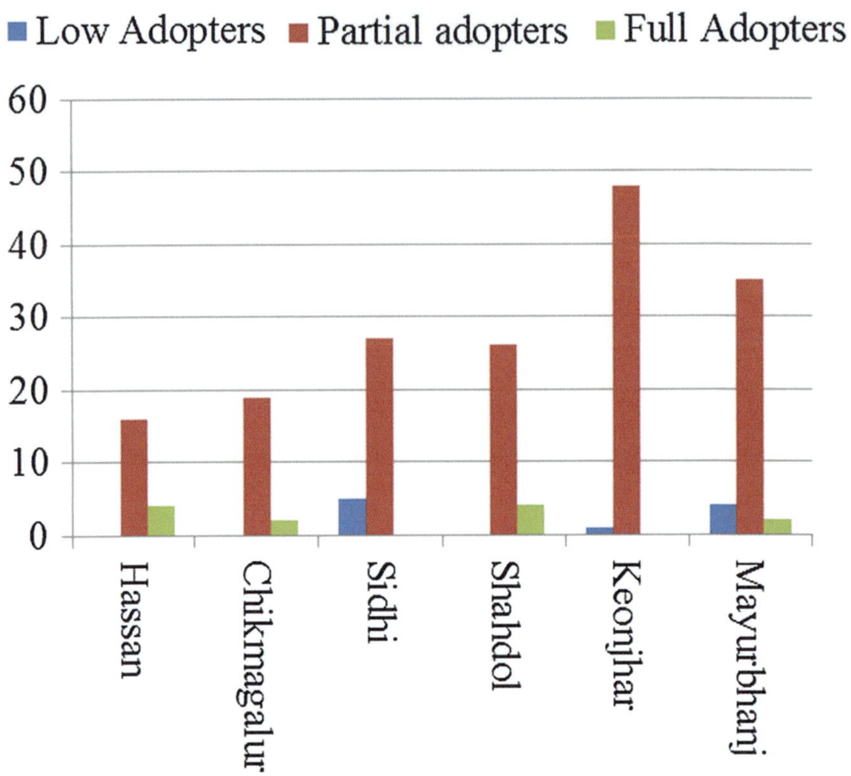

Fig. 6.4 Adoption level across districts. *Source* Survey data

Table 6.1 Adoption of SRI practices by farmers (in %)

Package	Hassan	Chikmagalur	Sidhi	Shahdol	Keonjhar	Mayurbhanj
Young seedling	95	100	37.5	83.33	79.59	97.56
Shallow planting	80	76.19	96.88	100	97.96	95.12
Seedling	20	23.81	75	83.33	85.71	70.73
Wider spacing	100	90.48	65.63	66.67	81.63	65.85
Use of organics	100	57.14	65.63	23.33	6.12	7.32
Use of cono-weeder	95	100	6.25	63.33	67.35	46.34
Alternate wetting and drying	55	38.1	28.13	93.33	67.35	48.78

Source Survey data

few in Keonjhar whereas there were no low adopters of SRI in the districts of Hassan, Chikmagalur and Shahdol (see Fig. 6.2).

Also, analysis of adoption of SRI practices across different districts showed that there were huge disparities in the level of adoption. The variations were noted even within the same state. For example when 63% of SRI farmers used cono-weeder for weeding in Shahdol, only 6% of SRI farmers used cono-weeder for weeding in Sidhi. Similarly, when 93% of SRI farmers practiced alternate wetting and drying in Shahdol, only 28% of SRI farmers did so in Sidhi. In another case, when 83% of SRI farmers used young seedling, only 37.5% SRI farmers in Sidhi used young seedling. The disparities in adoption were the highest in Madhya Pradesh followed by Karnataka. In Hassan, all farmers used organics for SRI cultivation whereas in Chikmagalur only 57% of SRI farmers used organics (see Table 6.1).

6.3 Socio-economic Profile—An Overview

Household survey was conducted in six districts drawn from three states. A random sample of SRI adopters and non-adopters from each district was selected. The total number of households interviewed was 386. The total sample consisted of equal number of adopters and non-adopters. Agriculture was the main occupation and livelihood strategy for most of the farm households in the study districts. Farming was the main occupation for around 80% of total households interviewed. The farmers with farming as main occupation were slightly higher among adopters than non-adopters of SRI indicating greater possibility of full-time farmers choosing adoption. Among the adopters, around 86% of farmers had farming as their main occupation, whereas among the non-adopters, around 74% had farming as their main occupation (see Fig. 6.5).

Majority of the farm households interviewed were either marginal farmers or small farmers. Marginal farmers were around 45%, small farmers around 36%, semi-medium farmers around 16%, medium farmers around 3% and large famers were less than 1% (see Fig. 6.6).

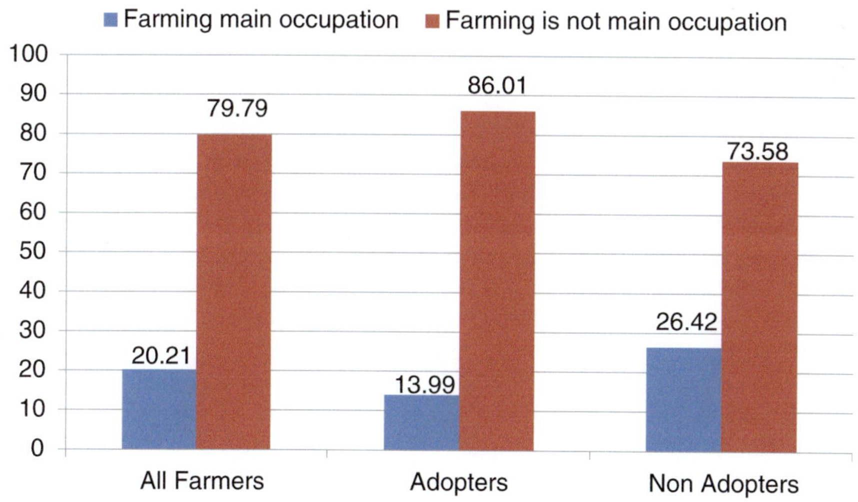

Fig. 6.5 Farmers' occupation. *Source* Survey data

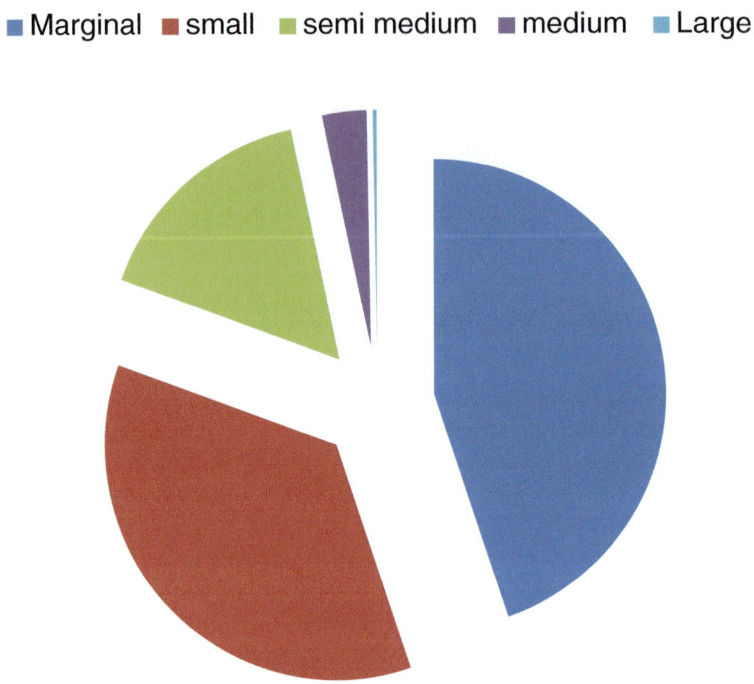

Fig. 6.6 Distribution of farmers according to farm size. *Source* Based on field survey data

Table 6.2 Gender of the head of the household

Gender	All		Adopters		Non-adopters	
	Nos.	%	Nos.	%	Nos.	%
Female	37	9.59	19	9.84	18	9.33
Male	349	90.41	174	90.16	175	90.67
Total	386	100	193	100	193	100

Source Survey data

Most farm households had male head of the households and the percentage share of male head of households among the total households was similar among both adopters and non-adopters (see Table 6.2).

Paradoxically, even with wider adoption of SRI, the percentage of non-adopters who did not have any information regarding SRI was the highest in Orissa (26%), followed by Madhya Pradesh (16%) and Karnataka (10%) (see Table 6.3). Around 43% of the households opined that they were facing difficulty in terms of availability of labourers. The percentage of farmers who expressed difficulty in getting labourers was more or less the same among adopters and non-adopters indicating that availability of labourers was a major concern. Perhaps this could be a reason for lower level of adoption of SRI in Karnataka even with relatively better information about SRI. The percentage of non-adopters who expressed difficulty in getting

Table 6.3 Sources of information about SRI

Sources	All		Adopters		Non-adopters	
	Nos.	%	Nos.	%	Nos.	%
No information about SRI	38	9.84	0	0	38	19.69
Agricultural department	139	36.01	90	46.63	48	24.87
NGO	155	40.16	96	49.74	59	30.57
Other	54	13.99	7	3.63	48	24.87
Total	386	100	193	100	193	100

Source Survey data

labourers was one of the highest in Karnataka (56%) and lowest in Madhya Pradesh (34%). In Orissa, among the non-adopters, around 56% of the farmers expressed difficulty in obtaining labourers.

Out of total farmers interviewed, only 38 farmers did not have any information regarding SRI. This constitutes only 10% of total sample and 20% of total non-adopters. The results indicate that lack of information may not be the main reason behind lack of adoption (see Table 6.3). Among the major sources of information regarding SRI, NGOs were the most proactive in terms of providing information about SRI to farmers. Around 40% of farmers received information about SRI from various NGOs. This was followed by agricultural department. Around 36% of farmers received information about SRI from government departments.

If the lack of adoption is not primarily due to lack of information, then it must be due to other hindrances such as lack of extension services, training. The data shows that, out of total rice farmers, only around 47% of farmers had access to any sort of extension services (see Table 6.3). The access to extension services was particularly limited among the non-adopters indicating that lack of extension services can be a reason for non-adoption. When around 57% of adopters had access to extension services, around 63% of non-adopters did not have any access to extension services (see Fig. 6.7).

As far as the labour availability is concerned, around 43% of total farmers reported that they faced difficulty in getting labourers for rice cultivation, with the difficulty faced by non-adopters being slightly higher than adopters of SRI. Among the adopters, only around 37% faced difficulty in getting labourers whereas among the non-adopters around 49% faced difficulty in getting labourers (see Fig. 6.8).

The data on source of irrigation for rice farmers shows that majority of the farm households relied solely on rainfall for irrigation. Around 50% of farmers had only rainfall as the main source for irrigation. After rainfall, the main source of irrigation was canal, followed by wells. Those who had borewell or sprinkler were only around 3%. The reliance on rainfall was higher among non-adopters than adopters. Among the adopters, only around 40% relied upon rainfall whereas among the non-adopters around 60% had only rainfall as the main source. This indicates that the problem of lack of irrigation facility can also be a reason for non-adoption. After

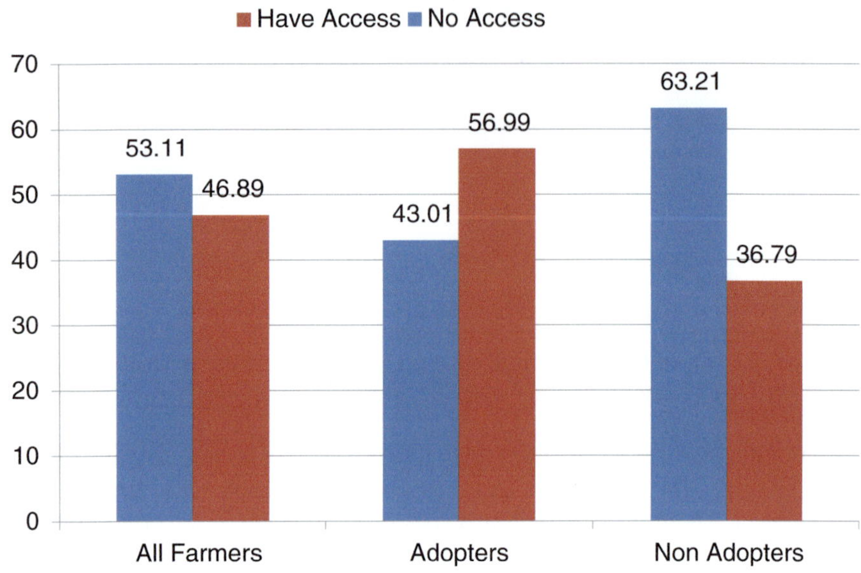

Fig. 6.7 Access to extension services. *Source* Survey data

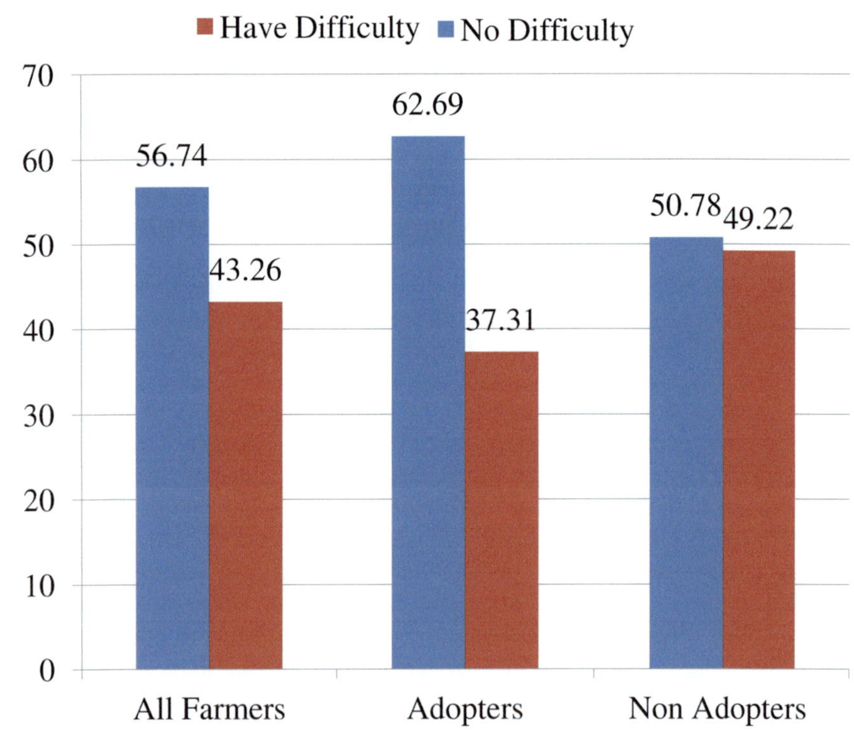

Fig. 6.8 Labour availability. *Source* Survey data

rainfall, canal and well remained the major sources for irrigation among both adopters as well as non-adopters (see Table 6.4).

As far as the membership in farmers' organisations is concerned, around 60% of farmers did not have any membership with input supply cooperatives. The percentage of farmers who did not have any membership was higher among the non-adopters than adopters (see Fig. 6.9).

Table 6.4 Source of irrigation

Source of irrigation	All		Adopters		Non-adopters	
	Nos.	%	Nos.	%	Nos.	%
Rainfall	192	49.74	77	39.90	116	60.10
Pond	5	1.30	5	2.59	0	0.00
Well	54	13.99	38	19.69	15	7.77
Borewell/sprinkler	13	3.37	10	5.18	3	1.55
Lake	2	0.52	1	0.52	1	0.52
River	38	9.84	20	10.36	18	9.33
Canal	82	21.24	42	21.76	40	20.73
Total	386	100	193	100	193	100

Source Survey data

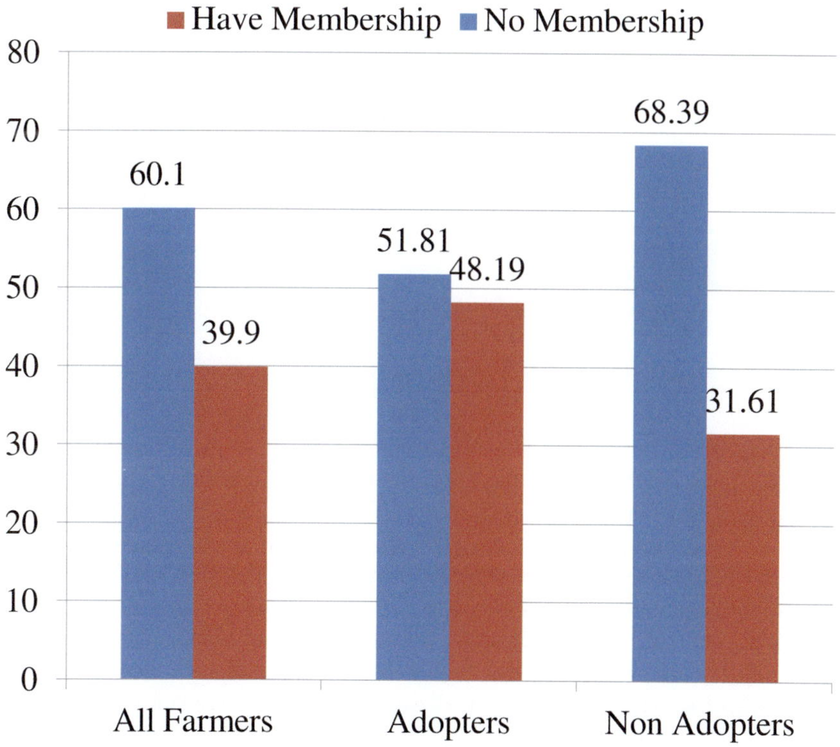

Fig. 6.9 Membership with input supply cooperatives. *Source* Survey data

Table 6.5 Source of income

Source of income	All Nos.	%	Adopters Nos.	%	Non-adopters Nos.	%
Farming	191	49.48	106	54.92	85	44.04
Service	1	0.26	0	0.00	1	0.52
Trading	2	0.52	0	0.00	2	1.04
Wage labour	3	0.78	0	0.00	3	1.55
Dairying/poultry	7	1.81	0	0.00	7	3.63
Farming and service	21	5.44	12	6.22	9	4.66
Farming and wage labour	126	32.64	56	29.02	70	36.27
Farming and dairying	14	3.63	9	4.66	5	2.59
Farming and other	21	5.44	10	5.18	11	5.70
Total	386	100	193	100	193	100

Source Survey data

Majority of the farm households had farming not only as their main occupation but also as main source of income (see Table 6.5). Around 50% of total households, 55% of total adopters and 44% of non-adopters had farming as their main source of income. This was followed by wage labour. Around 33% of total households, 29% of total adopters and 36% of total non-adopters had wage labour as main source of income. The percentage share of farmers who had farming as the main source of income was less among the non-adopters compared to adopters, while the percentage share of wage labour was relatively higher among non-adopters compared to adopters. This indicates that the lack of SRI adoption by non-adopters may not be due to availability of better opportunities but rather could be due to other factors such as lack of one's own cultivated land, higher rent for assets.

6.4 Socio-economic Profile—District-Wise

In Hassan district, 90% of farmers had farming as the main occupation leaving only 10% who did not have farming as their main occupation (see Table 6.6). However, none of the farmers in Hassan adopting SRI had farming as their main occupation, and all the farmers who did not have farming as the main occupation have adopted SRI. Among the non-adopters of SRI in Hassan, around 20% of farmers had farming as main occupation whereas 80% of farmers did not have farming as their main occupation. This shows that the adopters of SRI in Hassan are mainly those farmers who have other sources of income. The same was true in the case of other districts as well. In all other districts, the percentage of farmers who had other sources of income was relatively higher among the adopters than among the non-adopters (see Table 6.6). Among all the districts, Chikmagalur had the highest percentage of SRI farmers who had farming as main occupation. This was followed

Table 6.6 Farming as main occupation—district-wise

Districts	All farmers		Adopters		Non-Adopters	
	Yes (in %)	No (in %)	Yes (in %)	No (in %)	Yes (in %)	No (in %)
Chikmagalur	64.29	35.71	33.33	66.67	38.10	61.90
Hassan	90.00	10.00	0.00	100.00	20.00	80.00
Shahdol	83.33	16.67	13.33	86.67	20.00	80.00
Sidhi	71.88	28.13	21.88	78.13	34.38	65.63
Keonjhar	83.67	16.33	14.29	85.71	24.49	75.51
Mayurbhanj	81.71	18.29	12.20	87.80	24.39	75.61
All	79.79	20.21	15.54	84.46	26.42	73.58

Source Survey data

by Sidhi. So the data generally indicates that the adoption of SRI is the highest among those farmers who have other sources of income.

The district-wise analysis of distribution of farmers according to farm size shows that Mayurbhanj district had the highest percentage of marginal farmers (59.8%) in the sample. This was followed by Sidhi (56.25%). The marginal farmers were the lowest in Hassan (22.5%) followed by Chikmagalur (28.6%). Small farmers were the highest in Chikmagalur (42.8%), followed by Keonjhar (41.84%). Small farmers were the lowest in Sidhi (25%), followed by Hassan (32.50%) and Mayurbhanj (32.93%). Semi-medium farmers were the highest in Hassan (10%) followed by Shahdol (8.33%). The medium farmers were the highest in Hassan (35%) followed by Chikmagalur (23.81%). The medium farmers were the lowest in Mayurbhanj (7.32%), followed by Keonjhar (11. 22). Numbers of large farmers were relatively less in the sample (see Figs. 6.10, 6.11, 6.12, 6.13, 6.14 and 6.15).

As far as the sources of information about SRI across different districts are concerned, NGOs were the major source of information followed by agricultural departments. The striking observation was about Madhya Pradesh. In Sidhi, all the adopters of SRI came to know about SRI through NGOs, whereas in Shahdol all the adopters of SRI came to know about SRI through agricultural departments. The fact that Shahdol is a NFSM district could be the reason for active involvement of the agricultural departments in spreading the awareness about SRI. Similarly, in Chikmagalur, 86% of adopters of SRI came to know about SRI through NGOs, whereas in Hassan 85% of farmers came to know about SRI through agricultural departments (see Table 6.7). Again Hassan is an NFSM district. The same holds true in the case of Madhya Pradesh as well. In Keonjhar, 73.5% of adopters came to know about SRI through agricultural departments whereas in Mayurbhanj 85.37% of adopters came to know about SRI through NGOs. This broadly indicates the involvement of government in spreading the awareness about SRI in NFSM districts and in non-NFSM districts the information gap is filled by NGOs.

Among the non-adopters, around 20% of farmers had no information about SRI. The percentage of non-adopters who did not have any information about SRI is therefore not very high. The involvement of government in spreading the awareness

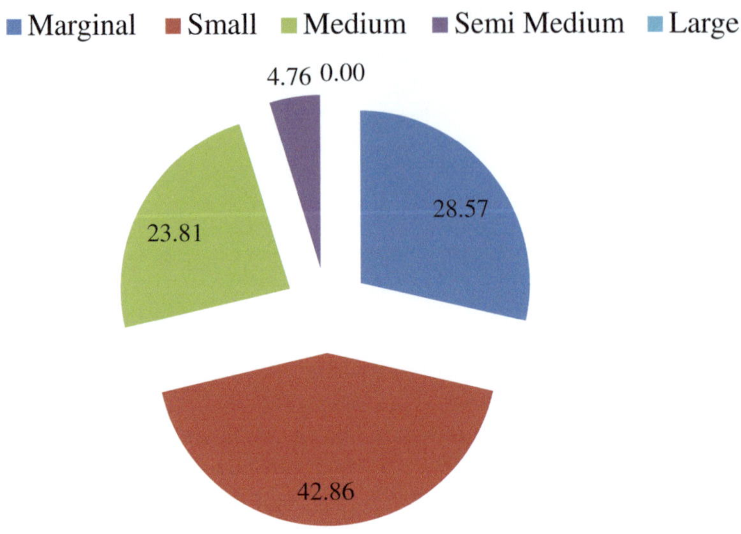

Fig. 6.10 Distribution of farmers according to farm size—Chikmagalur. *Source* Based on field survey data

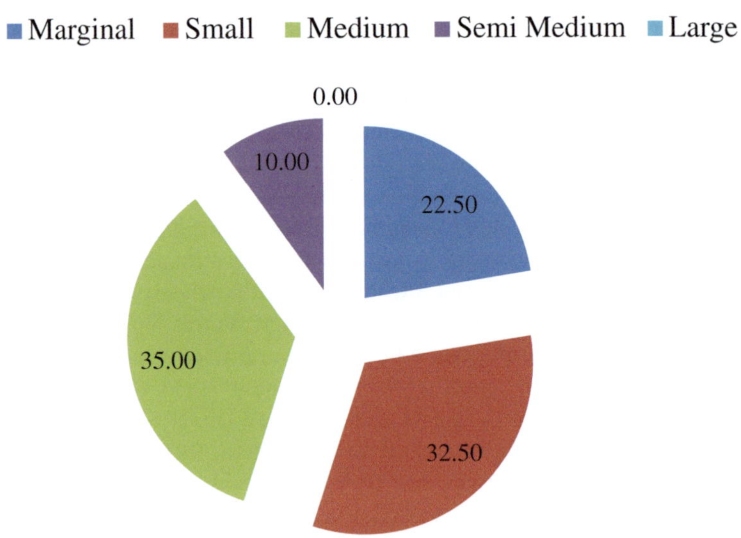

Fig. 6.11 Distribution of farmers according to farm size—Hassan. *Source* Based on field survey data

among the farmers was the highest in Shahdol (a NFSM district) and the lowest in Sidhi (a non-NFSM district). NGOs were very active among the non-adopters of Mayurbhanj (non-NFSM district). Despite having the information about NGOs and

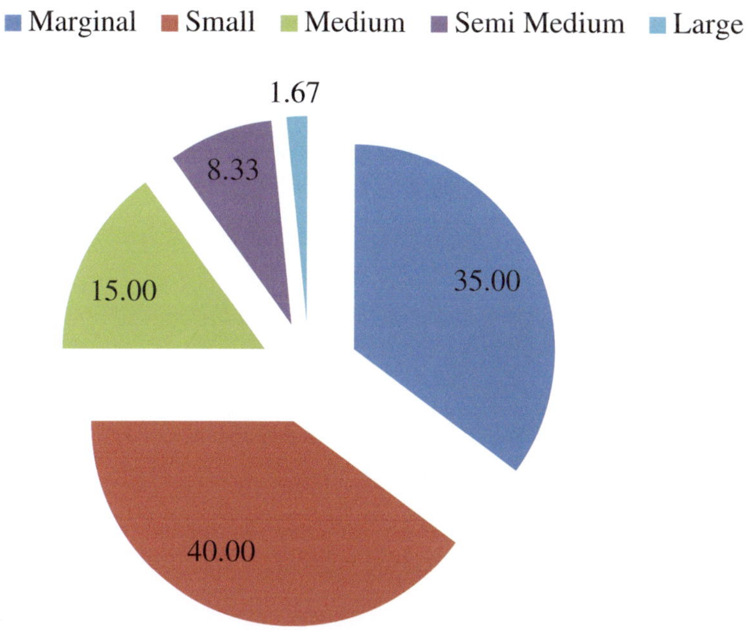

Fig. 6.12 Distribution of farmers according to farm size—Shahdol. *Source* Based on field survey data

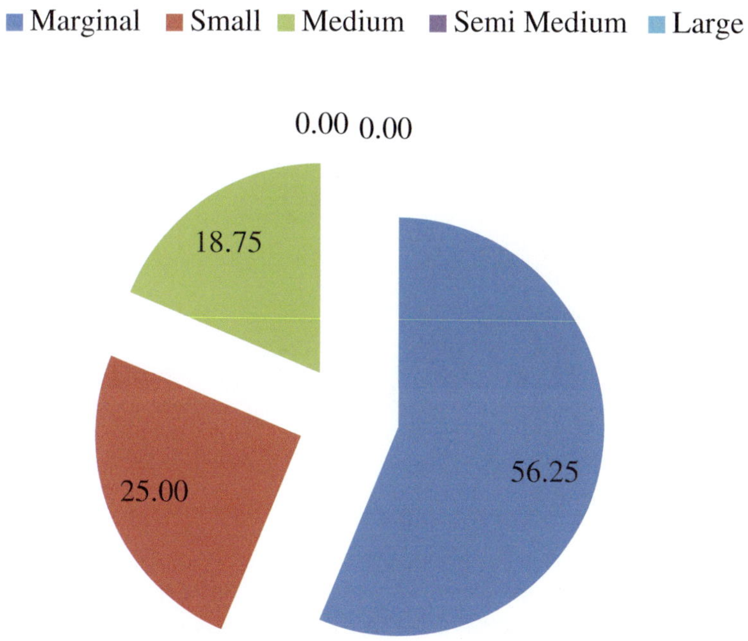

Fig. 6.13 Distribution of farmers according to farm size—Sidhi. *Source* Based on field survey data

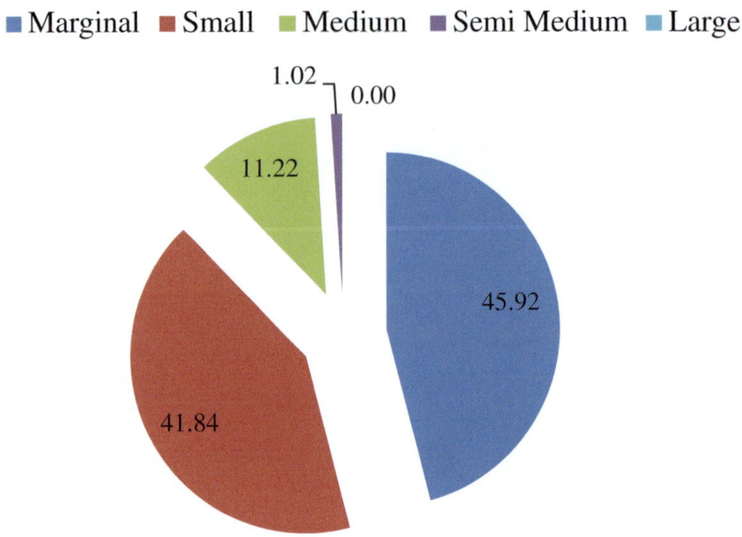

Fig. 6.14 Distribution of farmers according to farm size—Keonjhar. *Source* Based on field survey data

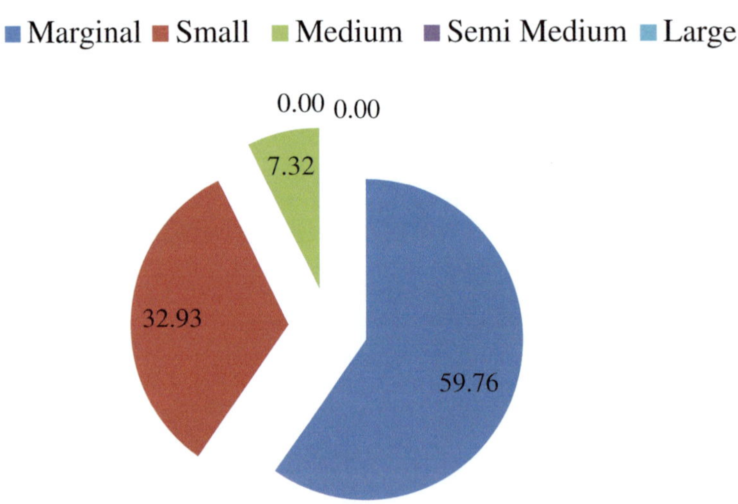

Fig. 6.15 Distribution of farmers according to farm size—Mayurbhanj. *Source* Based on field survey data

strong government involvement in spreading the awareness about SRI, there is not much adoption of SRI taking place in districts such as Shahdol and Hassan. This shows that there are other factors that affect the adoption of SRI.

Table 6.7 Sources of information—district-wise

District	Adopters (in %)				Non-adopters (in %)			
	1	2	3	4	1	2	3	4
Chikmagalur		14.29	85.71	0.00	14.29	19.05	52.38	14.29
Hassan		85.00	5.00	10.00	5.00	45.00	0.00	50.00
Shahdol		100.00	0.00	0.00	13.33	60.00	0.00	26.67
Sidhi		0.00	100.00	0.00	18.75	0.00	37.50	43.75
Keonjhar		73.47	20.41	6.12	22.45	30.61	24.49	22.45
Mayurbhanj		9.76	85.37	4.88	31.71	4.88	60.98	2.44
All		46.63	49.74	3.63	19.69	24.87	31.09	24.35

1 No information about SRI, *2* Agricultural department, *3* NGOs, *4* Others
Source Survey data

Table 6.8 Access to extension services

District	Adopters		Non-adopters	
	Have access	No access	Have access	No access
Chikmagalur	33.33	66.67	33.33	66.67
Hassan	80.00	20.00	65.00	35.00
Shahdol	100.00	0.00	60.00	40.00
Sidhi	12.50	87.50	12.50	87.50
Keonjhar	26.53	73.47	71.43	28.57
Mayurbhanj	41.46	58.54	31.71	68.29
All	45.08	54.92	46.63	53.37

Source Survey data

The above findings show that 60% of non-adopters in Shahdol and 65% of non-adopters in Hassan had access to extension services. Access to extension services among the non-adopters was the highest in Keonjhar followed by Hassan and Shahdol (see Table 6.8). Interestingly, only 12.5% of adopters in Sidhi had access to extension services. Access to extension services among the adopters were the highest in Shahdol followed by Hassan. A striking observation is there was not much difference in access to extension services between the adopters and non-adopters, and in some cases, the lack of access was higher among the adopters than non-adopters. For example in Keonjhar, 73.47% of adopters lacked proper extension services, whereas only 28.57% of non-adopters lacked access to extension services. Even among the total number of farmers, 54.92% of adopters of SRI farmers had no access to extension services, whereas 46.6% of non-adopters had access to extension services.

Out of the total sample of farmers, around 57% of farmers did not face any difficulty in getting labourers. This was due to the greater availability of labourers in Sidhi district. The data showed that the labour availability was the highest in Sidhi followed by Mayurbhanj and Shahdol. In districts such as Sidhi, Mayurbhanj and Shahdol, fewer number of farmers reported the difficulty in getting labourers for rice cultivation (see Fig. 6.14); however, Keonjhar and Hassan reported great difficulty in getting labourers. The number of farmers who reported difficulty in getting labourers was, therefore, high in Keonjhar and Hassan, followed by Chikmagalur (see Fig. 6.16).

Out of the total sample of farmers, around 45% of farmers did not have a proper irrigation facility (see Fig. 6.17). The data showed that around 45% of farmers were dependent on rainfall for irrigation. The greater difficulty in irrigation was noted in Sidhi district where around 44% of farmers were fully dependent on rainfall for irrigation. The other sources for irrigation were the highest in a relatively developed district—Hassan (see Fig. 4.17). This was followed by Chikmagalur and Shahdol.

The farmers who had no membership in input supply cooperatives were around 60% in the sample. Sidhi had the highest number of respondents who did not have any membership in input supply cooperatives (see Fig. 6.18). This was followed by Mayurbhanj and Hassan. The highest number of membership was noted in Shahdol. This was followed by Chikmagalur (see Fig. 6.18).

The highest percentage of respondents who had farming as the only source of income was observed in Shahdol (68.33%). This was followed by Sidhi (51.56%) and Keonjhar (50%) (see Table 6.9). The farmers with farming and wage labour as the source of income was the highest in Mayurbhanj (39.02%), followed by Hassan

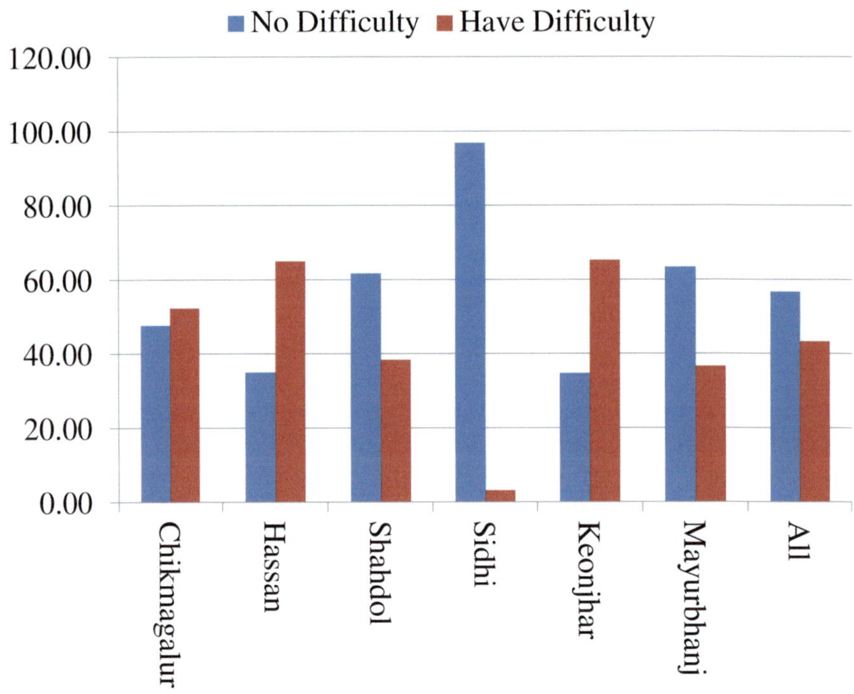

Fig. 6.16 Labour availability for rice cultivation—district-wise. *Source* Survey data

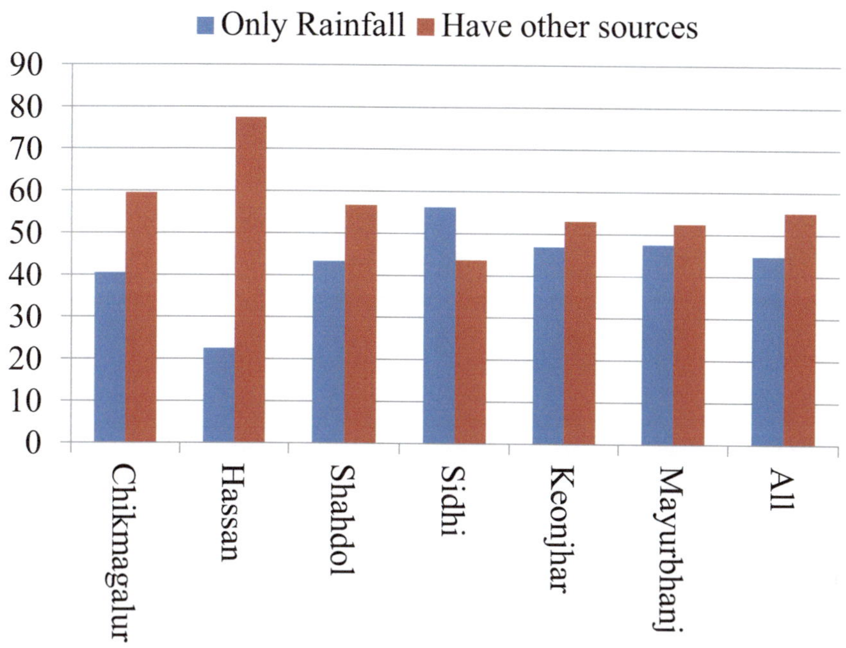

Fig. 6.17 Access to irrigation—district-wise. *Note* Other sources = Pond, well, borewell/sprinkler, lake, river, canal, etc. *Source* Survey data

(35%) and Sidhi (34.38%) (see Table 4.8). The farmers with farming and dairying were the highest in Hassan (25%). The source of income as farming and dairying was negligible among other respondents in other districts. Farming and other sources of income was the highest in Chikmagalur (see Table 6.8). This data generally indicates that the income from off-farm activities was generally higher among the respondents from Karnataka state.

6.5 Economics of SRI Adoption—Rice Yield, Household Farm Income and Cost of Cultivation

The descriptive analysis of rice showed that adopters of SRI received higher average yield as compared to non-adopters. The average rice yield for adopters of SRI was 3.43 quintal per acre whereas for the non-adopters of SRI the average yield was only 2.7 quintal per acre (see Table 6.10). Similarly, the average farm income was also higher for the adopters as compared to non-adopters. The average income for adopters was 80 thousand whereas for non-adopters the average income was only 75 thousand. As far as the cost of rice cultivation is concerned, the data shows that the average cost of cultivation was lower for adopters than non-adopters. The

Fig. 6.18 Membership with input supply cooperatives. *Source* Survey data

Table 6.9 Source of Income (in % of total sample farmers)

District	Farming	Service and trading	Wage labour	Farming and dairy/poultry	Farming and service	Farming and wage labour	Farming and other
Chikmagalur	47.62	0	0	0	7.14	30.95	14.29
Hassan	35	0	0	25	2.5	35	2.5
Shahdol	68.33	0	0	0	0	26.67	5
Sidhi	51.56	0	0	1.56	9.38	34.38	3.13
Keonjhar	50	3.06	3.06	5.1	4.08	29.59	5.1
Mayurbhanj	41.46	0	0	6.1	8.54	39.02	4.88

Source Survey data

average cost of cultivation for adopters was only Rs. 4257 per acre whereas for non-adopters it was around Rs. 7697 per acre (see Table 6.10).

The state-wise analysis of the yield, income and cost of cultivation also revealed that yield, cost and income benefits were higher among the adopters of SRI. The

Table 6.10 Descriptive statistics for economics of SRI adoption

Variable	Adopters	Non-adopters
Average yield (quintal per acre)	3.43 (1.09)	2.7 (0.89)
Average income (in lakhs rupees)	0.80 (0.29)	0.75 (0.28)
Average cost of cultivation (in rupees per acre)	4257.04 (3723.6)	7696.95 (9079.8)
Total observations	193	193

Note Standard deviation is given in parentheses
Source Survey data

Table 6.11 Descriptive statistics for economics of SRI adoption (state-wise)

Districts	SRI adoption status	Average yield (quintal per acre)	Average income (in lakhs rupees)	Average cost of cultivation (in rupees per acre)
Karnataka	Adopters	4.14 (0.72)	1.11 (0.37)	4390.8 (2991.03)
	Non-adopters	3.60 (0.74)	1.07 (0.39)	6709.3 (5937.3)
Madhya Pradesh	Adopters	2.99 (1.05)	0.76 (0.19)	4284.9 (3486.5)
	Non-adopters	2.00 (0.56)	0.68 (0.15)	9090.7 (8534.5)
Orissa	Adopters	3.40 (1.10)	0.68 (0.18)	4176.8 (4188.5)
	Non-adopters	2.77 (0.73)	0.660 (0.18)	7186.8 (10,489.2)

Note Standard deviation is given in parentheses
Source Survey data

average yield among the adopters was the highest in Karnataka. The average yield in Karnataka among the adopters was 4.14 quintal per acre. The average yield among the adopters was the lowest in Madhya Pradesh. The average yield among the SRI adopters of Madhya Pradesh was only 2.99 quintal per hectare (see Table 6.11). Interestingly, even the non-adopters of SRI in Karnataka had a relatively higher rice yield as compared to even the adopters of SRI in other two states. This shows the greater disparity in productivity across states.

As was the case with yield benefits, the income benefits were also higher in Karnataka. The average income among the adopters was higher than non-adopters in all the states. In Karnataka, the average income among the adopters was around Rs. 1.11 lakhs whereas among the non-adopters the average income was around Rs. 1.07 lakhs. It must be noted that even the non-adopters had higher average income in Karnataka as compared to adopters in the other two states. The difference in income between adopters and non-adopters was the highest in Madhya Pradesh and lowest in Orissa (see Table 6.11).

Average cost of rice cultivation was lower for adopters of SRI in all states. The cost of cultivation among the SRI-adopted farmers was not considerably different across states although Karnataka had relatively higher cost. The average cost of cultivation among the adopters of SRI in Karnataka, Madhya Pradesh and Orissa was Rs. 4390 per acre, Rs. 4285 per acre and Rs. 4177 per acre, respectively. Among the non-adopters, the cost was the highest in Madhya Pradesh. In Madhya

Pradesh, the per acre cost of cultivation for rice was Rs. 9091. Therefore, the difference in cost of cultivation between the adopters and non-adopters were also the highest in Madhya Pradesh. The difference was the lowest in Karnataka (see Table 6.11).

A further analysis at the district level also showed that yield, income and cost of cultivation gains were higher among the adopters of SRI than non-adopters of SRI. In all the six districts, yield and income were higher and cost of cultivation was lower for adopters of SRI (see Table 6.12). The average rice yield experienced by the adopters of SRI was much higher in Karnataka. There have not been many differences in average rice yield between Chikmagalur and Hassan although the average yield in Hassan was marginally higher. The differences in average rice yield among the adopters were the highest in Madhya Pradesh. In Shahdol, adopters of SRI experienced average rice yield of 3.74 quintal per acre whereas in Sidhi the adopters of SRI experienced only average rice yield of 2.29 quintal per acre. The differences in average cost of cultivation were the highest in Shahdol. In Shahdol, the adopters of SRI incurred an average cost of Rs. 3489 per acre whereas non-adopters incurred Rs. 9399 per acre. The average cost of cultivation among the adopters was also the lowest in Shahdol (see Table 6.12). These point out the urgent need for region-specific government interventions and programmes for enhancing SRI dissemination and adoption.

SRI consists of three main principles—plant management, soil management and water management (Chap. 9 discusses them in detail). The average yield from the adoption of all these three principles was higher than the non-adoption of practice as well as the non-adoption of SRI (see Table 6.13). The loss from yield benefits was relatively lower in the case of plant management indicating that the yield gap

Table 6.12 Descriptive statistics for economics of SRI adoption (district-wise)

District	SRI adoption status	Average yield (quintal per acre)	Average income (in lakhs rupees)	Average cost of cultivation (in rupees per acre)
Chikmagalur	Adopters	4.14 (0.83)	1.05 (0.43)	4500.56 (2891.7)
	Non-adopters	3.26 (0.49)	0.97 (0.29)	7266.98 (6126.2)
Hassan	Adopters	4.15 (0.60)	1.27 (0.40)	4275.5 (3163.1)
	Non-adopters	4.00 (0.80)	1.08 (0.37)	6123.7 (5831.2)
Shahdol	Adopters	3.74 (0.86)	0.87 (0.21)	3489.5 (3183.6)
	Non-adopters	2.33 (0.45)	0.74 (0.16)	9398.8 (9045.6)
Sidhi	Adopters	2.29 (0.65)	0.66 (0.09)	5030.5 (3640.3)
	Non-adopters	1.70 (0.49)	0.62 (0.11)	8801.8 (8161.2)
Keonjhar	Adopters	3.70 (1.15)	0.71 (0.21)	4657.4 (4559.1)
	Non-adopters	3.0 (0.73)	0.67 (0.20)	7107.3 (7977.3)
Mayurbhanj	Adopters	3.05 (0.94)	0.65 (0.13)	3602.8 (3670.4)
	Non-adopters	2.49 (0.62)	0.64 (0.15)	7281.7 (12,977.7)

Note Standard deviation is given in parentheses
Source Survey data

Table 6.13 Average yield (per quintal) for each component of SRI adoption

SRI Practice/component	Adopters of practice	Non-adopters of practice	Non-adopters of SRI
Plant management	3.59 (1.09)	3.02 (1.00)	2.7 (0.89)
Soil management	3.43 (1.09)	2.7 (0.89)	2.7 (0.89)
Water management	3.77 (1.06)	2.99 (0.97)	2.7 (0.89)

Note Standard deviation is given in parentheses
Source Survey data

Table 6.14 Average yield (per quintal) for the mutually exclusive combinations of SRI adoption

SRI practice/component	Average yield (per quintal)	Number of observations
Non-adopters	2.7 (0.89)	193
Adopters of plant plus water management only	3.69 (0.80)	19
Adopters of plant plus soil management only	3.23 (0.95)	47
Adopters of plant plus water plus soil management	3.79 (1.11)	89

Note Standard deviation is given in parentheses
Source Survey data

was substantially higher from the adoption of soil and water management principles.

Similarly, the adoption of single or combination of these principles provided substantial yield gains as compared to non-adoption (see Table 6.14). It may be noted that the average yield was substantially higher under the full adoption than partial adoption. For example, the average yield was 3.79 quintal per acre as compared to the adoption of plant plus water management (3.69 quintal per acre) or plant plus soil management (3.23 quintal per acre) (see Table 6.14).

Analysis of yield benefits under the individual practices showed that generally the average yield benefits were higher than non-adoption of SRI as well as the non-adoption of each individual practices. The yield gap was very high in the case of adoption of young seedling. The average yield per quintal was only 2.49 for non-adopters of young seedling whereas for adopters of young seedling the average yield per quintal was 3.65. This indicates the importance of the component of young seedling.

Interestingly, the average yield was lower for the adoption of shallow planting and single seedling than the non-adoption of these practices (Table 6.15).

The average income was higher for the adopters of all principles of SRI than non-adopters. The difference in income was higher for the adoption of plant management. The average income for adopters of plant management was around 83 thousand whereas for non-adopters of plant management the average income was around 71 thousand (see Table 6.16).

As was the case with yield benefits, the income benefits were also higher under the full adoption of SRI. For example, the average income for the adopters of all

Table 6.15 Average yield (per quintal) for individual SRI practices/combinations

SRI practice/component	Adopters of practice	Non-adopters of practice	Non-adopters of SRI
Young seedling	3.65 (1.01)	2.49 (0.92)	2.7 (0.89)
Shallow planting	3.42 (1.10)	3.47 (0.93)	2.7 (0.89)
Single seedling	3.34 (1.08)	3.60 (1.09)	2.7 (0.89)
Wider spacing	3.51 (1.05)	3.15 (1.18)	2.7 (0.89)
Use of organics	3.51 (0.96)	3.38 (1.15)	2.7 (0.89)
Use of cono-weeder	3.78 (1.05)	2.95 (0.96)	2.7 (0.89)
Alternate wetting and drying	3.77 (1.06)	2.99 (0.97)	2.7 (0.89)

Note Standard deviation is given in parentheses
Source Survey data

Table 6.16 Average income (in lakhs rupees) for each component of SRI adoption

SRI practice/component	Adopters of practice	Non-adopters of practice	Non-adopters of SRI
Plant management	0.83 (0.31)	0.71 (0.22)	0.75 (0.28)
Soil management	0.80 (0.29)	0.75 (0.28)	0.75 (0.28)
Water management	0.84 (0.33)	0.74 (0.21)	0.75 (0.28)

Note Standard deviation is given in parentheses
Source Survey data

Table 6.17 Average income (in Lakhs rupees) for the mutually exclusive combinations of SRI practices

SRI practice/component	Average yield (per quintal)	Number of observations
Non-adopters	0.75 (0.28)	193
Adopters of plant plus water management only	0.79 (0.32)	19
Adopters of plant plus soil management only	0.80 (0.25)	47
Adopters of plant plus water plus soil management	0.86 (0.33)	89

Note Standard deviation is given in parentheses
Source Survey data

principles was 86 thousand, which was higher than the adopters of other combinations as well as non-adopters of SRI (see Table 6.17). Following full adoption was the adoption of plant plus soil management which showed higher income benefits than other combinations.

As was the case with the analysis of yield for individual practices, the average income was lower for the adopters of shallow planting and young seedling as compared to non-adopters of these practices. The adopter of shallow planting and single seedling experienced an average income of 79 thousand and 77 thousand as compared to 87 thousand and 86 thousand, respectively, for non-adopters. This shows that the adoption of these two practices leads neither to yield benefits nor income benefits. Except for these two practices, the average income was higher for the adopters of all other practices compared to the non-adopters of any of the practices (see Table 6.18). The differences in average income were the highest for the adoption of cono-weeder (see Table 6.18).

The analysis of average cost of cultivation between the adopters and non-adopters of each SRI principle showed that the average cost for the adopters of plant management was lower than the non-adopters of plant management (see Table 6.19). The average cost for adopters of plant management was Rs. 4257 per acre, and for non-adopters of plant management, the average cost was Rs. 7697 per acre. Similarly, the average cost for adopters of water management was also lower than the non-adopters of water management. The average cost for adopters of water management was Rs. 3724 per acre whereas the same was around Rs. 4934 per acre for the non-adopters of water management. However, it may be noted that the average cost was higher for the adopters of soil management as compared to the non-adopters of soil management. The adopters of soil management incurred an average cost of Rs. 5023 per acre whereas the non-adopters of soil management incurred only Rs. 2429 per acre (see Table 6.19).

The analysis for the adoption of combinations of SRI principles shows that the average cost was substantially lower for the full adopters, whereas for the adopters

Table 6.18 Average income (in lakhs rupees) for individual SRI practices/combinations

SRI practice/component	Adopters of practice	Non-adopters of practice	Non-adopters of SRI
Young seedling	0.82 (0.30)	0.72 (0.22)	0.75 (0.28)
Shallow planting	0.79 (0.28)	0.87 (0.36)	0.75 (0.28)
Single seedling	0.77 (0.28)	0.86 (0.30)	0.75 (0.28)
Wider spacing	0.82 (0.31)	0.72 (0.20)	0.75 (0.28)
Use of organics	0.93 (0.36)	0.73 (0.21)	0.75 (0.28)
Use of cono-weeder	0.86 (0.32)	0.72 (0.22)	0.75 (0.28)
Alternate wetting and drying	0.84 (0.33)	0.74 (0.21)	0.75 (0.28)

Note Standard deviation is given in parentheses
Source Survey data

Table 6.19 Average cost of cultivation (rupees per acre) for each component of SRI adoption

SRI Practice/component	Adopters of practice	Non-adopters of practice	Non-adopters of SRI
Plant management	4257.05 (3723.6)	7696.9 (9079.8)	7696.9 (9079.8)
Soil management	5023.3 (3961.8)	2428.85 (2207.2)	7696.9 (9079.8)
Water management	3724.5 (3507.8)	4933.7 (3897.4)	7696.9 (9079.8)

Note Standard deviation is given in parentheses
Source Survey data

of plant plus soil management practices the cost was very high. The adopters of all principles incurred the average costs of Rs. 3973 per acre whereas the adopters of plant plus soil management incurred an average cost of Rs. 7013 per acre (Table 6.20).

The analysis of average cost differences under the adoption of individual practices showed that the average cost was higher for the adoption of 4 out of seven practices. The average cost for the adopters of young seedling, wider spacing, use of organics and use of cono-weeder was higher than for the non-adopters of these practices. The cost differences were the highest for the use of cono-weeder followed by the use of organics (see Table 6.21). This supports the results from the analysis done for different principles. It can be more costly to adopt the soil management due to the cost of organics and the cono-weeders.

Table 6.20 Average cost of cultivation (rupees per acre) for the mutually exclusive combinations of SRI practices

SRI practice/component	Average cost of production (rupees per acre)	Number of observations
Non-adopters	7696.9 (9079.8)	193
Adopters of plant plus water management only	2562.3 (3126.2)	19
Adopters of plant plus soil management only	7012.8 (3970.3)	47
Adopters of plant plus water plus soil management	3972.6 (3550.8)	89

Note Standard deviation is given in parentheses
Source Survey data

Table 6.21 Average cost of cultivation (rupees per acre) for individual SRI practices/combinations

SRI practice/component	**Adopters of practice**	**Non-adopters of practice**	**Non-adopters of SRI**
Young seedling	4295.9 (3794.2)	4093.2 (3453.9)	7696.9 (9079.8)
Shallow planting	4252.4 (3808.3)	4321.6 (2345.2)	7696.9 (9079.8)
Single seedling	3893.5 (3669.3)	4989.7 (3752.9)	7696.9 (9079.8)
Wider spacing	4278.7 (3784.9)	4187.9 (3559.9)	7696.9 (9079.8)
Use of organics	4885.4 (3597.6)	3930.5 (3759.8)	7696.9 (9079.8)
Use of cono-weeder	4908.2 (3993.9)	3375.6 (3136.1)	7696.9 (9079.8)
Alternate wetting and drying	3724.5 (3507.8)	4933.66 (3897.4)	7696.9 (9079.8)

Note Standard deviation is given in parentheses
Source Survey data

6.6 Adoption of SRI—Rice Varieties in the Study Region

The analysis of the SRI varieties adopted in various study districts confirms the view that SRI is not specific to any particular seed variety. In most of the districts, SRI farmers had used the same variety that they use for traditional rice cultivation. The number of varieties used by farmers was also found in abundance in these regions. In Chikmagalur district of Karnataka, the popular SRI varieties were Intan and Tunga. But largely farmers cultivate rajamudi rice for their own consumption due to preference for taste. In Hassan district of the same state, the popular SRI varieties were GK and Tunga. In both of these districts in Karnataka, the traditional rice variety—rajamudi—that farmers mainly use for their own consumption was also used by some farmers for SRI cultivation. In Shahdol district of Madhya Pradesh, the popular varieties were IR64, KPH 199 and DRRH-3. There were variations in varieties seen in the same state. For example in Sidhi, although IR64 was among the popular SRI variety, the other varieties that the farmers cultivated were GK, champion and other local varieties. The adoption of local varieties for SRI cultivation was one of the highest in Sidhi district. In Keonjhar district of Orissa, the popular SRI varieties were lolatto and 27P31, and in Mayurbhanj district, the popular SRI varieties were pooja, pratikaya, sahabhagi and suvarna. The varieties adopted by SRI farmers in the study region strengthens the view that SRI is not specific to any particular seed variety and can be cultivated using any rice variety depending upon the availability and other regional characteristics.

6.7 Conclusion

This chapter provided an overview of the adoption of SRI across different states and discussed the socio-economic profile of the sample households. The preliminary analysis of the data undertaken in this chapter revealed that majority of farm households were poor or marginal with farming or wage labour as their main sources of income. Additionally, majority of the households interviewed were headed by males. Interestingly, lack of information about SRI was not a major issue as only 10% of farmers lacked information regarding SRI. On the other hand, 53% of total farmers and 63% of non-adopters did not have access to extension services clearly indicating that lack of training and guidance in terms of adoption was big concerns. Non-availability of labour was reported by most farmers as a major concern even among adopters of SRI. Similarly, 40% of adopters and 60% of non-adopters solely relied upon rainfall as the major source of irrigation indicating that lack of irrigation was a likely reason for non-adoption. There were huge variations in socio-economic profiles of farmers across different districts as well. The analysis also showed that in NFSM districts, the main source of information about SRI was the government agricultural departments, whereas in non-NFSM districts the main source of information about SRI was NGOs. This shows that SRI is taking roots in India through diverse channels.

This chapter identified the socio-economic profile of the farm households along with certain constraints such as lack of access to extension services, irrigation facility, availability of labour. The analysis based on the descriptive statistics of yield, household income and cost of cultivation showed that the adopters of SRI had higher yield, income and cost benefits. Although the analysis showed that adoption of some components like soil management is more costly than others, the adoption of SRI is generally beneficial for rice farmers. The remaining chapters will be devoted to more rigorous analysis to strengthen our arguments.

The next chapter will deal with econometric estimation of the impact of several socio-economic factors on adoption decision by explicitly considering the effective information constraints in adoption.

Chapter 7
Depth and Intensity of SRI Adoption

7.1 Introduction

This chapter attempts to analyse factors influencing the intensity (in terms of acres devoted for SRI) and the depth (in terms of number of SRI components adopted) of adoption of SRI by explicitly considering the effective information constraints in the adoption decision. This chapter develops two types of simultaneous equation model, with the first set of simultaneous equation models analysing the status of SRI adoption (yes/no) in the first stage and the intensity (depth) of SRI adoption in the second stage. The second set of simultaneous equation model incorporates effective information constraint in the first stage and the intensity (depth) of SRI adoption in the second stage.

Despite all the efforts to promote SRI in India, the pace of spread of SRI adoption is slow in India (Nirmala et al. 2015). Studies have pointed out various constraints faced by farmers in adopting SRI (Moser and Barrett 2003; Doberman 2004; Noltze et al. 2012; Barrett et al. 2016). The importance of information on adoption is widely recognised in the technology adoption literature in general (Marra et al. 2001; Matuschke and Qaim 2009; Kabunga et al. 2012a, b; Lidia et al. 2012; Genius et al. 2013; Shiferaw et al. 2015; Singh et al. 2015) and in SRI literature in particular (Sigdel et al. 2014; Barrett et al. 2016). Just as limited access to information is a constraint in adoption, so also awareness of technology is a necessary condition for its adoption (Diagne and Demont 2007).

When there are constraints in accessing a new technology or information related to a new method, even farmers with positive demand for adoption may not be able to adopt due to multiple constraints in adoption (Shiferaw et al. 2015). Nonetheless, studies investigating adoption of agricultural technologies in the context of binding constraints are very limited. In fact, many adoption decision studies assume that farmers function in a perfect information setting and, therefore, face an unconstrained access to technology. According to Shiferaw et al. (2015), under such conditions of the zero (non-adoption) generating process, an adoption decision is

P. Varma, *Rice Productivity and Food Security in India*,
DOI 10.1007/978-981-10-3692-7_7

modelled using probit and logit models for non-divisible technologies and Tobit type models for divisible technologies.

Even in a perfect information setting, farmers with positive desired demand for adopting a new technology may fail to realise this potential demand owing to various constraints (Croppenstedt et al. 2003; Shiferaw et al. 2008, 2015). The relaxation of constraints may lead to an increased adoption of new technology and, therefore, modelling technology adoption by dividing farmers into adopters and non-adopters fail to bring out the difference between actual and desired demand (Shiferaw et al. 2015). This may lead to inconsistency in estimated parameters.

Kabunga et al. (2012a, b) analysed the adoption of tissue culture (TC) bananas in Kenya, which is a relatively knowledge-intensive technology. Although many farmers are aware of TC, they lacked the skill in adopting the TC successfully. As a result, the adoption rates remain relatively low. SRI is also a knowledge-intensive method. Therefore, information constraints play an important role. In our study, we define information as "effective information". The importance of the effectiveness of information has also been highlighted in the literature (Marra et al. 2001; Lidia et al. 2012; Singh et al. 2015).

The present chapter makes an attempt to model the factors affecting the adoption of SRI conditional upon access to effective information. Around 90% of households in our sample did have information about SRI (see Appendix Table 7.7). However, only 56% of non-adopters received information about SRI from agricultural departments or NGOs, when around 96% of adopters had information about SRI from agricultural departments or NGOs. Therefore, the information received by around 25% of non-adopters was from other sources such as neighbours and farmers. This shows that information in itself was not a reason for non-adoption; rather, the effectiveness of information was more important.

In the context of SRI, there is evidence that farmers are constrained by information and skills necessary for local adaptation of the SRI methods (Sigdel et al. 2014; Barrett et al. 2016). Also, where government agencies have gotten involved in SRI dissemination or where NGOs have had sufficient resources for their extension activities, the spread has been quite rapid (Uphoff and Kassam 2009). As mentioned already, SRI is a knowledge-intensive cultivation technique, and therefore, it requires local adaptation of the methods with proper guidance.

Most of the earlier econometric models applied for studying the agricultural technology adoption behaviour relied largely on a probit or logit type models if the data is collected on the basis of whether the technology is adopted or not. If the researchers have the data on the level of adoption, the censored regression model, also called the Tobit model is commonly used. The underlying assumption of Tobit type specification is that the farmer with positive desired demand for adoption will have an unconstrained access to technology (Shiferaw et al. 2008). The Tobit type specification has no mechanism to categorise households with positive desired demand into the households with constraints and households without any constraints. As a result, the households with constraints but a positive desired demand for adoption will be automatically treated as non-adopters in such specifications. The non-adopters with positive desired demand for adoption may turn out to be adopters when the constraints are relaxed. Therefore, modelling technology adoption by dividing farmers into adopters and non-adopters fails to bring out the difference between actual and desired demand (Shiferaw et al. 2015). This may lead to inconsistency in estimated parameters.

Therefore, in this chapter, we develop a simultaneous equation model by considering effective information as a partially binding constraint in the adoption of SRI. Several studies have applied double-hurdle models (Noltze et al. 2012), the variants of double-hurdle models (Croppenstedt et al. 2003) and multiple-hurdle models (Shiferaw et al. 2008, 2015; Asfaw et al. 2010).

There are a few studies that analyse the adoption using average treatment effects framework as well by differentiating technology exposure (awareness) from adoption (Diagne and Demont 2007; Kathage et al. 2016). The study by Kathage et al. (2016) examined whether information is an important constraint to hybrid maize adoption in Tanzania, along with other factors that explain the adoption behaviour.

However, there are hardly any such studies, in the context of SRI, explicitly specifying a constraint in adoption. An understanding of the constraints and other factors influencing adoption of environment-friendly methods such as SRI is pertinent to overcome the hurdles that farmers face in the process. In line with the methodological approach used by Croppenstedt et al. (2003) and Shiferaw et al. (2008, 2015), this study analyses the role of effective information in conditioning SRI adoption by rice farmers in selected states of India. Although the study focuses mainly on SRI adoption, it intends to make a contribution to agricultural technology adoption in general and SRI adoption literature in particular. The joint probability for adoption is estimated using conditional (recursive) mixed process estimator (CMP) developed by Roodman (2009, 2011).

The remainder of this chapter is organised as follows: Section 7.2 provides conceptual framework for household technology adoption in the presence of constraints along with variable description and hypothesised relationships. Section 7.3 deals with the model specification, and the main analytical results are presented and discussed in Sect. 7.4. Concluding observations and policy implications is presented in Sect. 7.5.

7.2 Conceptual Framework

The farmer's decision on whether to adopt a new technology or not is based on utility maximisation (Rahm and Huffman 1984; Shiferaw et al. 2015). The *i*th farmer will go for a new technology if the utility derived from the new technology (U_{1i}) is greater than the old technology (U_{0i}), i.e. $U_{1i} > U_{0i}$. By denoting A_{d} for adoption decision, we can write:

$$A_{\mathrm{d}} = \begin{cases} 1 & \text{if } U_{0i} < U_{1i} \\ 0 & \text{if } U_{1i} \leq U_{0i} \end{cases} \tag{7.1}$$

In the first scenario ($A_{\mathrm{d}} = 1$), the utility from the new technology is higher, whereas in the second scenario ($A_{\mathrm{d}} = 0$) the utility is smaller than or equal to the old technology. The probability that the farmer adopts a superior technology ($A_{\mathrm{d}} = 1$) depends on a set of explanatory variables.

$$\begin{aligned} P_i &= P_r(A_{\mathrm{d}} = 1) = P_r(U_{1i} > U_{0i}) \\ &= P_r\left[\alpha_1 F_i(R_i Y_i) + e_{1i} > \alpha_0 F_i(R_i Y_i) + e_{o,i}\right] \\ &= P_r\left[e_{1i} - e_{0,i} > F_i(R_i, L_i)(\alpha_1 - \alpha_0)\right] \\ &= P_r(\mu_i > \ - F_i(R_i L_i)\beta) \\ &= F_i(X_i \beta) \end{aligned} \tag{7.2}$$

where X is the $n \times k$ matrix of the explanatory variables and β is the $k \times 1$ vector of parameters to be estimated, Pr(.) is the probability function, μ_i is the random error term and $F_i(X_i\beta)$ is the cumulative distribution function for μ_i evaluated at $X_i\beta$. The probability that a farmer will adopt a superior method is a function of the vector of explanatory variables and of the unknown parameters and error term.

As discussed earlier, the expected utility of the new technology is not, however, the only one factor that determines the adoption. This is especially true for small holder farmers in developing countries where they face information constraints in adoption. The information, i, that is required for a farmer to make the adoption decision can be given as:

$$A_i = \begin{cases} 1 & \text{if } A_i > 0 \\ 0 & \text{if } A_i \leq 0 \end{cases} \tag{7.3}$$

Now the farmer is aware of the new method, and the information is received from the agricultural departments or NGOs. Passing the first hurdle places a farmer in the class of farmers who are "potential adopters" due to the fact they have "effective information". This is expected to help a farmer in evaluating the benefits of the new method. Whether the new method has been adopted or not by the households can be given as:

$$A = A_i A_{\mathrm{d}} = \begin{cases} 1, & \text{if the new method is adopted} \\ 0, & \text{if the new method is not adopted} \end{cases} \tag{7.4}$$

The decision to adopt depends on their household and farm-level characteristics along with other factors. If they adopt the SRI, we say that they have crossed the second hurdle.

7.3 Model Specification

The farmer's demand for new method can be written as below:

$$y_i^* = \alpha \boldsymbol{x}_i' + u_i \tag{7.5}$$

where $\boldsymbol{x}_i'$ is vector of variables that determine the demand function, α is a parameter vector, u is an error term with mean 0 and variance σ_u. Similarly, the latent variable underlying a farmer's access to effective information can be modelled as below:

$$I_i^* = \beta z_i' + \epsilon_i (\text{Access to effective information}) \tag{7.6}$$

In the above equations, z_i' is the vector of variables that affect the availability of information. And β is the parameters to be estimated; ϵ is the error terms with mean 0 and variance 1. The observed demand for SRI method by a farmer (Y_i) is characterised by the interaction of models (7.5) and (7.6).

The joint probability for adoption is estimated using conditional (recursive) mixed process estimator (CMP) developed by Roodman (2009, 2011).[1] CMP estimates multiequation, recursive mixed process models. "Mixed process" means that different equations can have different kinds of dependent variables. CMP can only fit "recursive" models with clearly defined stages. To illustrate, A and B can be determinants of C, and C a determinant of D—but D cannot be a determinant of A, B or C (Roodman 2011). Equation (7.6) is estimated using probit models, while a truncated normal model estimates the intensity of adoption (Eq. 7.5, where Y_i represents the adoption of farmer in terms of the intensity of adoption).

The intensity of SRI adoption is defined as the number of SRI acres devoted by household under rice acres. The depth of adoption is defined in terms of the number of SRI practices adopted by a household[2]. In the depth of adoption model, the first equation is estimated using the probit type specification and the last equation for the

---

[1]The model is estimated using CMP command in Stata as it helps us to run multiple simultaneous equations.

[2]The depth of adoption of SRI is analysed using the total practices as dependent variable. The SRI emerged as a set of six practices. They are as follows:

1. Transplanting of young seedlings
2. Shallow planting of seedlings
3. Single seedling at wider spacing
4. Weeding by mechanical weeder
5. Use of organics
6. Efficient water management: Alternate wetting and drying

For the purpose of our analysis, we have decided to split the third practice into two—single seedling and wider spacing. We observed that many farmers, although allowing wider spacing, were planting more than one seedling.

depth of adoption is modelled using an ordered probit specification. The ordered probit analyses the factors that influence the adoption of a combination of SRI practices (in terms of total number of practices adopted). The variables that affect the adoption of a SRI practice may differently affect the depth of adoption of practices (Teklewold et al. 2013). The probability for adopting first practice can differ from the probability of adopting a second practice or a third practice and so on. The farmers who adopt greater number of practices are definitely superior in terms of the depth of adoption.

7.3.1 Description of Variables

Several studies have included household and farm characteristics as important factors influencing the adoption decision by farmers (Feder et al. 1985; Uaiene 2011; Teklewold et al. 2013; Ogada et al. 2014; Manda et al. 2015). As far as the age factor is concerned, one set of studies postulate a positive relationship (Meshram et al. 2012; Kassie et al. 2013) while the other a negative relationship (Manda et al. 2015). Those who postulate a positive relationship argue that older farmers are more experienced and might have accumulated greater physical and social capital (Kassie et al. 2013). Nonetheless, there is also belief that older farmers are less amenable to change and, therefore, unwilling to change from old practices to new ones (Adesina and Zinnah 1993). There are studies that have found positive relationship between age and adoption decisions (Langyintuo and Mungoma 2008; Meshram et al. 2012), while negative relationship between the two has also been noted by researchers (Teklewold et al. 2013).

Certain fixed social bias (i.e. gender of household head) is also expected to have an impact on technology adoption (Langyintuo and Mungoma 2008). There is a view that women farmers face greater constraints in terms of access to resources and time and hence can be less enthusiastic in adoption (Pender and Gebremedhin 2008). Manda et al. (2015) found a negative and significant relationship between the gender of the household and adoption of improved maize varieties. However, Doss and Morris (2001) argue that there is no significant association between the gender of the farmer and the probability of adoption.

The size of the household is used as a proxy to capture labour endowment (Pender and Gebremedhin 2008). As far as the importance of total farm size is concerned, studies on SRI show a positive relationship between the size of the farm and intensity of SRI in Timor Leste (Noltze et al. 2012), whereas studies on the adoption of improved maize varieties in Zambia showed a negative relationship between the two (Langyintuo and Mungoma 2008).

Education of the household is also expected to have a positive impact on adoption decisions (Moser and Barrett 2003; Pender and Gebremedhin 2008; Langyintuo and Mungoma 2008; Meshram et al. 2012). The study done by Haldar et al. (2012) for Bardhaman district in West Bengal showed that education level is important in adopting SRI. Similarly, the study by Manda et al. (2015) observed a positive relationship between education and the adoption of sustainable agricultural practices.

Assets owned by the farmer are expected to relax many of the above-mentioned constraints and, therefore, are an important factor in deciding the adoption (Langyintuo and Mungoma 2008). Also, we consider the impact of farmers who have farming as main occupation on adoption decisions. The study by Noltze et al. (2013) shows that household heads whose main occupation is farming are much less likely to adopt SRI.

Access to off-farm activities and income in general is expected to have a positive impact on adoption decisions (Davis et al. 2009). The study by Langyintuo and Mungoma (2008) found a positive relationship in the case of adoption of improved maize varieties, and the study by Noltze et al. (2012) found a positive relationship in the case of SRI adoption. However, the studies by Mathenge et al. (2015) and Manda et al. (2015) found a negative relationship between the two. The negative relationship is also not surprising as greater access to off-farm activities can lead to a reduction in farm activities and investment in farm.

Similarly, certain components of SRI such as intermittent irrigation, although perceived to be water saving, require proper crop management and irrigation availability (Dobermann 2004). Therefore, there are studies that highlight the importance of irrigation and irrigation management in deciding adoption of SRI (Tsujimoto et al. 2009; Noltze et al. 2012; Uphoff 2012). Significant differences in adoption intensity between regions were noticed by some studies due to the plot specific characteristics (Langyintuo and Mungoma 2008). Around 70% of farmers in our sample did not have terrain which was levelled. Therefore, studies have found terrain type to be important in deciding adoption of SRI (Moser and Barrett 2003).

Membership in farmer organisation is another important factor that has been highlighted in the literature in affecting the adoption decisions (Barrett et al. 2004; Matuschke and Qaim 2009; Khonje et al. 2015). Market access also has a huge bearing on transaction cost in accessing information and any new technology or improved practices (Kassie et al. 2015). In line with Kassie et al. (2015), we consider the distance from main market as a proxy for market access. Access to extension services also plays a significant role in adoption decisions (Langyintuo and Mungoma 2008; Olwande et al. 2009; Mazvimavi and Twomlow 2009; Manda et al. 2015; Shiferaw et al. 2015). Since SRI is a knowledge-based innovation, extension services play an even greater role in wider adoption (Noltze et al. 2012). Studies in the context of technology adoption in general have confirmed this view (Langyintuo and Mungoma 2008).

As far as labour is concerned, there is no consensus in the literature with one set of studies arguing SRI as labour intensive (Moser and Barrett 2003; Senthilkumar et al. 2008; Noltze et al. 2012; Palanisami et al. 2013) and another set of studies arguing labour requirement is high only in the initial phase of adoption (Barrett et al. 2004; Uphoff 2012). The present study addresses a slightly different concern than analysing the issue of the amount of labour required for adoption of SRI. Our field experience showed that most farmers were concerned about the difficulty in finding labourers for rice cultivation in general and for SRI in particular. Several farmers including the adopted farmers highlighted this as a major constraint while adopting SRI. Therefore, the present study included a variable "availability of labourers" in our model. Additionally, our study makes use of the wage rates for male and female labourers to understand how the cost of labourers affects the adoption decisions.

Aversion to risk by farmers has also been highlighted by some studies as the reason for poor adoption (Johnson and Vijayaraghavan 2011). The present study has included the variable "crop failure" that the farmers experienced in the last five years as a proxy to capture the impact of risk-related factors on the adoption of SRI. Finally, in order to capture the impact of NFSM in promoting the adoption of SRI, we give dummy variable equal to one for those districts where SRI is incorporated within NFSM.

The intensity and the depth of adoption of SRI is modelled conditional upon access to effective information as the major constraint in adoption. The key variables hypothesised to affect *access to information* include household and farm characteristics such as age, gender, education, number of farm assets owned and farm size; social capital variables such as membership in farmer organisations and seed exchange experience; communication technology (ICT) such as mobile, access to off-farm activity and distance to main market. NFSM dummy is also incorporated to see whether NFSM has any differential impact in accessing information regarding SRI. A detailed description of variables is given in Table 7.1.

Table 7.1 Variable definitions

Variable	Definition
Age of the HoH	Age of the head of the household
Gender of HoH	Gender of the head of the household, dummy variable = 1 if the household has a male head of the household
Education	Number of family members who have education higher than tenth class
Household size	Number of the family members in the household including children
Farm size	Total size of owned and rented landholdings cultivated by household in hectares
Membership in farmer organisations	Membership of any of the family member in farmer organisations, dummy variable = 1 if any of the family member has membership in farmer organisations, = 0 otherwise
Distance to extension services	Distance from the household to nearest extension service provider in kilometres
ICT (Mobile)	Dummy variable = 1 if any of the household member has a mobile phone, = 0 otherwise
Distance from main market	Distance to the nearest main market in kilometres

(continued)

Table 7.1 (continued)

Variable	Definition
Seed exchange experience	Experience of exchanging seeds among the farmers
Wage rate for male	Male wage rate for rice cultivation in rupees per day
Wage rate for female	Female wage rate for rice cultivation in rupees per day
Irrigation facility	Dummy variable = 1 if the household had irrigation sources other than rainfall, = 0 otherwise
Assets owned	Total number of the type of farm assets (bullocks, tractors, machinery, etc.) owned by the household
Terrain type	Dummy variable = 1 if the terrain is levelled, = 0 otherwise
Cultivation of only rice	Dummy variable = 1 if the household cultivates only rice, = 0 otherwise
Access to off-farm activity	Dummy variable = 1 if the household had access to off-farm activity, = 0 otherwise
Farming main occupation	Dummy variable = 1 if the household's main occupation is farming, = 0 otherwise
Labour availability	Dummy variable = 1 if the household does not find any difficulty in getting labour for SRI cultivation, = 0 otherwise
Crop failure	Dummy variable = 1 if the household has experienced any type of crop failure in the last 5 years, = 0 otherwise
SRI information	Dummy variable = 1 if the household had information regarding SRI from agricultural departments and NGOs, = 0 otherwise
NFSM dummy	Dummy variable = 1 for those districts where SRI is incorporated within NFSM, = 0 otherwise

7.4 Estimation Results and Discussion

7.4.1 Descriptive Statistics

Descriptive statistics of the explanatory variables that are hypothesised to influence the constraints and the adoption of SRI are presented in Table 7.2. Among non-adopters around 38 farmers did not have any information about SRI. This is interesting as it shows that around 80% of non-adopted farmers were aware of SRI. Therefore, the lack of information might not have played a major role in non-adoption. Rather, it points to the importance of effective information. When around 96% of adopters received information from either agricultural departments or NGOs, only 55% of non-adopters received information from agricultural departments or NGOs. Access to off-farm activity was higher among the non-adopters than adopters. Around 38% of non-adopters had access to off-farm activities, whereas only 26% of adopters had access to off-farm activities. Availability of irrigation facility was relatively lower among the non-adopters (40%) than adopters (60%). When 48% of adopters had membership in farmer organisations, only 32% of non-adopters had membership in farmer organisations.

Table 7.2 Descriptive statistics for variables used in the model

Variables	Adopters	Non-adopters
Age of the HoH	50.24 (11.20)	50.08 (11.53)
Gender of HoH	0.10 (0.30)	0.09 (0.29)
Membership in farmer organisation	0.48 (0.50)	0.32 (0.47)
Household size	5.23 (2.06)	4.97 (1.77)
Education	0.90 (1.15)	0.68 (1.14)
Farm size	3.45 (3.02)	3.04 (2.66)
Seed exchange experience	0.64 (0.48)	0.57 (0.50)
ICT (Radio and TV)	0.31 (0.47)	0.28 (0.45)
ICT (Mobile)	0.70 (0.46)	0.71 (0.46)
Access to off-farm activity	0.26 (0.44)	0.38 (0.49)
Assets owned	2.01 (2.57)	0.97 (0.72)
Cultivation of only rice	0.27 (0.44)	0.42 (0.49)
Wage rate for male	167.10 (69.80)	167.93 (80.98)
Wage rate for female	136.40 (50.22)	139.90 (68.10)
Terrain type	0.32 (0.47)	0.26 (0.44)
Irrigation facility	0.60 (0.49)	0.40 (0.49)
Distance to extension service	11.34 (8.5)	12.29 (7.7)
Distance from main market (in km)	12.97 (13.42)	9.10 (6.32)
SRI information	0.96 (0.19)	0.56 (0.50)
Crop failure	0.23 (0.42)	0.33 (0.47)
Farming as main occupation	0.86 (0.35)	0.74 (0.44)
Labour availability	0.63 (0.48)	0.51 (0.50)
NFSM dummy	0.52 (0.50)	0.52 (0.50)
Number of observations	193	193

Note Standard deviation is given in parentheses

For 86% of adopters, farming was the main occupation whereas around 74% of non-adopters had farming as main occupation. Terrain was levelled for around 32% of adopters, whereas the same was levelled only for 26% of non-adopters. Around 42% of non-adopters had the cultivation of only rice, whereas around 27% of the adopters had the cultivation of only rice. Average distance to extension service was higher for non-adopters than adopters. However, the average distance to the nearest main market was higher for adopters than for adopters. When 63% of adopters had availability of labour for rice cultivation, only 51% of non-adopters had availability of labour for rice cultivation. The average wage rate also varied between women and men labourers. The average wage rate for women labourers among the adopters was Rs. 136, whereas for male labourers the average wage rate was around Rs. 167. The average number of farm assets owned by the adopters was also higher than by the non-adopters.

7.4.2 Status and Intensity of SRI Adoption (in Terms of Acres of Land Under SRI)

The results from the first set of simultaneous equation models on the status and intensity of adoption showed that farm assets had a positive and statistically significant impact on the adoption of SRI (see Table 7.3). Cultivation of only rice had a negative and statistically significant impact on adoption indicating farmers who cultivate only rice did not eagerly adopt SRI. This is intuitively true. The farmers who cultivate only rice may be poor with fewer resources available. The results also showed that better irrigation facility encouraged the adoption of SRI. The irrigation facility had a positive and statistically significant impact on adoption of SRI. The results substantiate our field experiences. Several farmers had opined lack of irrigation facility as one of the major reasons for non-adoption. Additionally, labour difficulty was also highlighted as one of the major obstacles. As expected, the availability of labourers had a positive and statistically significant impact on adoption of SRI. Similarly, effective information was statistically significant and

Table 7.3 Simultaneous equation model for adoption status and intensity of adoption (Rice acres under SRI)

	Coefficient	Z-statistic	Marginal effects
A. Adoption of SRI			
Age of the HoH	−0.01 (0.01)	−1.32	−0.00 (0.00)
Gender of HoH	0.25 (0.26)	0.94	0.07 (0.07)
Household size	0.05 (0.05)	1.04	0.01 (0.01)
Education	−0.03 (0.07)	−0.45	−0.01 (0.02)
Farm size	−0.01 (0.03)	−0.42	−0.00 (0.01)
Terrain type	−0.06 (0.21)	−0.30	−0.02 (0.06)
Assets owned	0.18 (0.06)	2.99***	0.05 (0.02)***
Cultivation of only rice	−0.34 (0.17)	−2.01**	−0.09 (0.05)**
Access to off-farm activity	−0.04 (0.25)	−0.16	−0.01 (0.07)
Irrigation facility	0.33 (0.16)	2.08**	0.09 (0.04)**
Membership in farmer organisation	0.26 (0.17)	1.53	0.07 (0.05)
Wage rate for female	−0.01 (0.00)	−1.75*	−0.00 (0.00)*
Wage rate for male	0.01 (0.00)	2.06**	0.00 (0.00)**
Distance from main market (in km)	0.19 (0.12)	1.60	0.05 (0.03)
Distance to extension service	−0.22 (0.14)	−1.58	−0.06 (0.04)
Labour availability	0.63 (0.18)	3.51***	0.18 (0.05)***
Farming main occupation	0.25 (0.30)	0.83	0.07 (0.08)
SRI information	1.68 (0.23)	7.24***	0.47 (0.05)***
NFSM dummy	−0.06 (0.17)	−0.36	−0.02 (0.05)
Constant	−1.90 (0.69)	−2.76***	

(continued)

Table 7.3 (continued)

	Coefficient	Z-statistic	Marginal effects
B. Intensity of SRI adoption			
Age of the HoH	−0.00 (0.00)	−1.97**	−0.00 (0.00)**
Gender of HoH	−0.05 (0.05)	−0.94	−0.02 (0.02)
Household size	0.00 (0.01)	0.15	0.00 (0.00)
Education	0.01 (0.01)	0.51	0.00 (0.01)
Farm size	0.00 (0.01)	0.12	0.00 (0.00)
Terrain type	0.01 (0.04)	0.36	0.01 (0.02)
Assets owned	0.01 (0.01)	1.40	0.00 (0.00)
Cultivation of only rice	−0.04 (0.03)	−1.29	−0.02 (0.01)
Access to off-farm activity	−0.04 (0.04)	−0.82	−0.01 (0.02)
Irrigation facility	0.03 (0.03)	1.14	0.01 (0.01)
Membership in farmer organisation	0.07 (0.03)	2.33**	0.03 (0.01)**
Wage rate for female	−0.00 (0.00)	−1.27	−0.00 (0.00)
Wage rate for male	0.00 (0.00)	0.92	0.00 (0.00)
Distance from main market (in km)	0.03 (0.02)	1.30	0.01 (0.01)
Distance to extension service	−0.03 (0.03)	−1.09	−0.01 (0.01)
Labour availability	0.08 (0.03)	2.67***	0.03 (0.01)***
Farming main occupation	−0.06 (0.05)	−1.26	−0.03 (0.02)
SRI information	0.19 (0.03)	5.65***	0.07 (0.01)***
NFSM dummy	0.00 (0.03)	0.21	0.00 (0.01)
Constant	0.18 (0.12)	1.49	
No. of observations	386	LR chi2(38)	157.89
Log likelihood	−129.66683	Prob > chi2	0.0000

Note *, ** and *** indicate significance levels at 10, 5 and 1% level, respectively. Standard errors are in parentheses

positive indicating the importance of effective information in encouraging the adoption of SRI.

Another interesting observation is the findings with respect to the wage rates. The female wage rate had a negative and statistically significant impact on adoption, whereas the male wage rates had a positive and statistically significant impact on adoption. These findings are quite interesting.

Majority of labourers employed in paddy cultivation in the selected districts were female labourers. The weeding operations under conventional rice cultivation have been traditionally done by women. However, as a result of SRI adoption, rice farmers hire more and more of male labourers for mechanical weeding (Senthilkumar et al. 2008). Therefore, the shift from manual weeding to mechanical weeding resulted in greater demand for male labourers, leading to a positive relationship between male wage rates and SRI adoption. This also points out the skill-intensive nature of SRI adoption and gender-biased technical change.

The results from the intensity adoption model showed that age of the head of the household had a negative impact on the intensity of adoption of SRI. The finding is consistent with the few studies on agricultural technology adoption (Adesina and Zinnah 1993; Teklewold et al. 2013; Manda et al. 2015). However, the finding is in sharp contrast to some studies that postulated a positive relationship between age and the technology adoption (Langyintuo and Mungoma 2008; Meshram et al. 2012; Kassie et al. 2013). Membership in farmer organisation, availability of labour and the access to effective information had a positive and statistically significant impact on the intensity of adoption.

The dummy variable NFSM was positive but insignificant in the intensity of adoption model indicating the NFSM did not have an expected significant impact on the intensity of adoption of SRI.

7.4.3 Access to Effective Information and the Intensity of Adoption (in Terms of Acres of Land Under SRI)

The results for second set of simultaneous equation model for the intensity of SRI adoption are presented in Table 7.4. The information access model results showed that the membership in farmer organisations increased the household access to information about SRI from government and NGOs. The relationship between membership in farmer organisation and the access to information was positive and statistically significant. This is in line with the findings of Shiferaw et al. (2015). Additionally, the experience of farmers in seed exchange experience also enhanced the access to information. The households who owned a mobile phone also had a

Table 7.4 Intensity of adoption under information constraints (Rice acres under SRI)

	Coefficient	Z-statistic	Marginal effects
A. Information access model			
Age of the HoH	0.00 (0.01)	0.42	0.00 (0.00)
Gender of HoH	0.02 (0.26)	0.07	0.01 (0.07)
Membership in farmer organisation	0.60 (0.17)	3.49***	0.17 (0.05)***
Household size	0.04 (0.05)	0.85	0.01 (0.01)
Education	0.05 (0.08)	0.67	0.02 (0.02)
Farm size	−0.00 (0.03)	−0.04	−0.00 (0.01)
Seed exchange experience	0.55 (0.16)	3.52***	0.15 (0.04)***
ICT (Radio & TV)	−0.13 (0.19)	−0.67	−0.04 (0.05)
ICT (Mobile)	0.30 (0.17)	1.80*	0.08 (0.05)*
Access to off-farm activity	−0.09 (0.23)	−0.38	−0.02 (0.06)
Assets owned	0.05 (0.05)	1.04	0.01 (0.01)
Distance from main market (in km)	0.03 (0.11)	0.30	0.01 (0.03)

(continued)

Table 7.4 (continued)

	Coefficient	Z-statistic	Marginal effects
Distance to extension service	−0.18 (0.13)	−1.42	−0.05 (0.04)
Farming main occupation	0.22 (0.26)	0.85	0.06 (0.07)
NFSM dummy	0.00 (0.16)	0.01	0.00 (0.04)
Constant	−0.26 (0.59)	−0.45	
B. Intensity of SRI adoption			
Age of the HoH	−0.00 (0.00)	−1.96**	−0.00 (0.00)**
Gender of HoH	−0.05 (0.05)	−0.94	−0.05 (0.05)
Household size	0.00 (0.01)	0.15	0.00 (0.01)
Education	0.01 (0.01)	0.51	0.01 (0.01)
Farm size	0.00 (0.01)	0.12	0.00 (0.01)
Terrain type	0.01 (0.04)	0.36	0.01 (0.04)
Assets owned	0.01 (0.01)	1.40	0.01 (0.01)
Cultivation of only rice	−0.04 (0.03)	−1.29	−0.04 (0.03)
Access to off-farm activity	−0.04 (0.04)	−0.82	−0.04 (0.04)
Irrigation facility	0.03 (0.03)	1.14	0.03 (0.03)
Membership in farmer organisation	0.07 (0.03)	2.33**	0.07 (0.03)**
Wage rate for female	−0.00 (0.00)	−1.25	−0.00 (0.00)
Wage rate for male	0.00 (0.00)	0.95	0.00 (0.00)
Distance from main market (in km)	0.03 (0.02)	1.30	0.03 (0.02)
Distance to extension service	−0.03 (0.03)	−1.09	−0.03 (0.03)
Labour availability	0.08 (0.03)	2.67***	0.08 (0.03)***
Farming main occupation	−0.06 (0.05)	−1.26	−0.06 (0.05)
SRI information	0.19 (0.03)	5.65***	0.19 (0.03)***
NFSM dummy	0.01 (0.03)	0.21	0.01 (0.03)
Constant	0.18 (0.12)	1.49	
No. of observations	386	LR chi2(33)	84.77
Log likelihood	−216.69835	Prob > chi2	0.0000

Note *, ** and *** indicate significance levels at 10, 5 and 1% level, respectively. Standard errors are in parentheses

greater likelihood of getting the access to effective information. The relationship between the variable mobile and the access to effective information was positive and statistically significant.

The dummy variable for NFSM was positive but statistically insignificant indicating that the access to effective information was not significantly different in those districts, where the SRI promotion was incorporated under the NFSM to increase production and productivity.

The results from the intensity adoption model were similar to the earlier model (see Sect. 7.4.2). Age of the head of the household had a negative and statistically significant impact on the intensity of SRI adoption, whereas the membership in farmer organisation, availability of labour, the access to effective information, etc., had a positive and statistically significant impact on the intensity of adoption.

The dummy variable NFSM was positive but insignificant in the intensity of adoption model indicating the NFSM did not have an expected significant impact on the intensity of adoption of SRI.

7.4.4 Status and Depth of SRI Adoption (in Terms of the Number of SRI Practices)

The first set of simultaneous equation hurdle models on the status and depth of adoption is presented in Table 7.5. The results for the status of adoption were similar to the earlier model (see Sect. 7.4.2). The factors influencing the depth of adoption of SRI showed that age of the head of the household had a negative and statistically significant impact on the depth of adoption. But more variables came out to be significant in the depth of adoption model as compared to the intensity of adoption. For example, assets owned by farmers were not significant in the intensity of adoption, whereas the same was positive and statistically significant in the case of the depth of adoption (see Table 7.5). The assets owned had a positive relationship with the intensity of adoption but the results were statistically insignificant. The finding is in line with the existing studies (Langyintuo and Mungoma 2008). Interestingly, the depth of adoption of SRI was lower among those households who

Table 7.5 Simultaneous equation model for depth of adoption (SRI practices)

	Coefficient	Z-statistic	Marginal effects
A. Adoption of SRI			
Age of the HoH	−0.01 (0.01)	−1.29	−0.00 (0.00)
Gender of HoH	0.25 (0.28)	0.89	0.07 (0.08)
Household size	0.05 (0.05)	1.01	0.01 (0.01)
Education	−0.03 (0.08)	−0.43	−0.01 (0.02)
Farm size	−0.01 (0.04)	−0.39	−0.00 (0.01)
Terrain type	−0.06 (0.23)	−0.28	−0.02 (0.06)
Assets owned	0.18 (0.07)	2.72***	0.05 (0.02)***
Cultivation of only rice	−0.34 (0.17)	−1.97**	−0.09 (0.05)**
Access to off-farm activity	−0.04 (0.26)	−0.15	−0.01 (0.07)
Irrigation facility	0.33 (0.17)	1.99**	0.09 (0.05)**
Membership in farmer organisation	0.26 (0.18)	1.46	0.07 (0.05)
Wage rate for female	−0.01 (0.00)	−1.72*	−0.00 (0.00)*
Wage rate for male	0.01 (0.00)	2.05**	0.00 (0.00)**
Distance from main market (in km)	0.19 (0.13)	1.49	0.05 (0.04)
Distance to extension service	−0.22 (0.15)	−1.51	−0.06 (0.04)

(continued)

Table 7.5 (continued)

	Coefficient	Z-statistic	Marginal effects
Labour availability	0.63 (0.19)	3.37***	0.17 (0.05)***
Farming main occupation	0.25 (0.30)	0.81	0.07 (0.08)
SRI information	1.68 (0.23)	7.16***	0.47 (0.05)***
NFSM dummy	−0.06 (0.18)	−0.34	−0.02 (0.05)
Constant	−1.90 (0.71)	−2.66***	
B. Depth of SRI adoption			
Age of the HoH	−0.01 (0.01)	−2.22**	−0.01 (0.01)**
Gender of HoH	0.27 (0.22)	1.23	0.27 (0.22)
Household size	0.04 (0.04)	1.08	0.04 (0.04)
Education	0.05 (0.06)	0.77	0.05 (0.06)
Farm size	0.03 (0.02)	1.03	0.03 (0.02)
Terrain type	−0.05 (0.17)	−0.29	−0.05 (0.17)
Assets owned	0.10 (0.03)	3.34***	0.10 (0.03)***
Other crop cultivated	−0.26 (0.14)	−1.84*	−0.26 (0.14)*
Access to off-farm activity	0.05 (0.19)	0.23	0.05 (0.19)
Irrigation facility	0.34 (0.13)	2.60***	0.34 (0.13)***
Membership in farmer organisation	0.43 (0.14)	3.09***	0.43 (0.14)***
Wage rate for female	−0.01 (0.00)	−2.94***	−0.01 (0.00)***
Wage rate for male	0.01 (0.00)	3.13***	0.01 (0.00)***
Distance from main market (in km)	0.07 (0.09)	0.82	0.07 (0.09)
Distance to extension service	−0.00 (0.11)	−0.02	−0.00 (0.11)
Labour availability	0.47 (0.14)	3.32***	0.47 (0.14)***
Farming main occupation	0.42 (0.24)	1.75*	0.42 (0.24)*
SRI information	1.27 (0.20)	6.51***	1.27 (0.20)***
NFSM dummy	0.09 (0.14)	0.63	0.09 (0.14)
No. of observations	386	LR chi2(38)	176.79
Log likelihood	−554.26093	Prob > chi2	0.0000

Note *, ** and *** indicate significance levels at 10, 5 and 1% level, respectively. Standard errors are in parentheses

were predominantly rice cultivators. The relationship between cultivation of only rice had a statistically significant and negative relationship with the depth of SRI adoption. Irrigation facility, membership in farmer organisations, availability of labourers, access to effective information, etc., had a positive and statistically significant impact on the depth of adoption of SRI.

Similar to the status of adoption of SRI, wage rates for male labourers had a positive impact whereas wage rates for female labourers had negative impact. The results further showed a positive and significant relationship between farmers with farming as main occupation and the depth of adoption of SRI. Nonetheless, the result contradicts the findings of Noltze et al. (2013) for Timor Leste. As per the study by Noltze et al. (2013), the household head whose main occupation is farming is much less likely to adopt SRI. The differences in results point out the fact that the nature, intensity and the factors contributing to it can vary quite considerably across regions.

The aversion to risk as expected had a negative impact on the depth of adoption. The relationship between crop failure and depth of SRI adoption was negative and statistically significant. The dummy variable NFSM was positive but insignificant in the depth of adoption model indicating that NFSM did not have an expected significant impact on the depth of adoption of SRI.

7.4.5 *Access to Effective Information and the Depth of Adoption (in Terms of Number of Practices)*

The model results for the depth of SRI adoption are presented in Table 7.6. The results for information access model were similar to the adoption intensity model in Table 7.4. The information access model results showed that the membership in farmer organisations increased the households' access to effective information. The relationship between membership in farmer organisation and the access to effective information was positive and statistically significant. Similarly, the experience of

Table 7.6 Depth of adoption under information constraints (Rice acres under SRI)

	Coefficient	Z-statistic	Marginal effects
A. Information access model			
Age of the HoH	0.00 (0.01)	0.43	0.00 (0.00)
Gender of HoH	0.02 (0.26)	0.07	0.01 (0.07)
Membership in farmer organisation	0.60 (0.18)	3.45***	0.17 (0.05)***
Household size	0.04 (0.05)	0.83	0.01 (0.01)
Education	0.05 (0.08)	0.67	0.02 (0.02)
Farm size	−0.00 (0.03)	−0.04	−0.00 (0.01)
Seed exchange experience	0.55 (0.16)	3.52***	0.15 (0.04)***
ICT (Radio & TV)	−0.13 (0.19)	−0.66	−0.04 (0.05)
ICT (Mobile)	0.30 (0.17)	1.71*	0.08 (0.05)*
Access to off-farm activity	−0.09 (0.23)	−0.38	−0.02 (0.07)
Assets owned	0.05 (0.05)	1.03	0.01 (0.01)
Distance from main market (in km)	0.03 (0.11)	0.30	0.01 (0.03)
Distance to extension service	−0.18 (0.13)	−1.40	−0.05 (0.04)
Farming main occupation	0.22 (0.26)	0.85	0.06 (0.07)
NFSM dummy	0.00 (0.16)	0.01	0.00 (0.04)
Constant	−0.26 (0.59)	−0.44	
B. Intensity of SRI adoption			
Age of the HoH	−0.01 (0.01)	−2.22**	−0.01 (0.01)**
Gender of HoH	0.27 (0.22)	1.23	0.27 (0.22)
Household size	0.04 (0.04)	1.08	0.04 (0.04)

(continued)

Table 7.6 (continued)

	Coefficient	Z-statistic	Marginal effects
Education	0.05 (0.06)	0.77	0.05 (0.06)
Farm size	0.03 (0.02)	1.03	0.03 (0.02)
Terrain type	−0.05 (0.17)	−0.29	−0.05 (0.17)
Assets owned	0.10 (0.03)	3.32***	0.10 (0.03)***
Other crop cultivated	−0.26 (0.14)	−1.85*	−0.26 (0.14)*
Access to off-farm activity	0.05 (0.20)	0.23	0.05 (0.20)
Irrigation facility	0.34 (0.13)	2.61***	0.34 (0.13)***
Membership in farmer organisation	0.43 (0.16)	2.76***	0.43 (0.16)***
Wage rate for female	−0.01 (0.00)	−3.00***	−0.01 (0.00)***
Wage rate for male	0.01 (0.00)	3.04***	0.01 (0.00)***
Distance from main market (in km)	0.07 (0.09)	0.82	0.07 (0.09)
Distance to extension service	−0.00 (0.11)	−0.02	−0.00 (0.11)
Labour availability	0.47 (0.14)	3.30***	0.47 (0.14)***
Farming main occupation	0.42 (0.24)	1.73*	0.42 (0.24)*
SRI information	1.27 (0.62)	2.04**	1.27 (0.62)**
NFSM dummy	0.09 (0.14)	0.63	0.09 (0.14)
No. of observations	386	LR chi2(34)	142.58
Log likelihood	−692.15893	Prob > chi2	0.0000

Note *, ** and *** indicate significance levels at 10, 5 and 1% level, respectively. Standard errors are in parenthesis

farmers in seed exchange experience also enhanced the access to effective information. The households who owned a mobile phone also increased the likelihood of getting the access to effective information. The relationship between the variable mobile and the access to information was positive and statistically significant. Similar to the intensity model, the variable NFSM was positive but statistically insignificant.

The results from the depth of adoption model showed that the results were similar to the depth of adoption model results using the first set of simultaneous equation hurdle model (see Table 7.5).

7.5 Conclusion

This chapter analysed the determinants of adoption of SRI by using two kinds of simultaneous equation models. The first set of simultaneous equation models analysed the status of SRI adoption (yes/no) in the first stage and intensity (depth) of SRI adoption in the second stage, whereas the second set of simultaneous equation models explicitly incorporated the effective information constraints in the model. When farmers lack effective information about a new practice, even farmers with a positive desired demand will not be able to adopt the new practice.

Therefore, in line with some of the recent studies (Croppenstedt et al. 2003; Asfaw et al. 2010; Shiferaw et al. 2008, 2015), the present chapter developed a simultaneous equation model by explicitly incorporating access to effective information as a constraint in adoption. There have not been any such attempts to analyse the information constraint in the context of SRI, which is especially true for India.

The results from the analysis offer unique policy-relevant insights into the importance of effective information in conditioning the intensity and the depth of SRI adoption in India. Even when around 90% of the sample farmers had information about SRI, only 76% of sample farmers received information about SRI from formal sources such as agricultural departments and NGOs. The argument is further strengthened by the fact that among the adopters around 96% of farmers received information from agricultural departments and NGOs, whereas among the non-adopters only 55% received information from agricultural departments and NGOs. The effective information is, therefore, deemed to be important in influencing the final adoption decision of farmers.

The information access model showed that none of the household or farm-level variables were statistically significant. Although variables such as education, size of the landholding, household size and farm assets owned showed positive sign, none of these variables were statistically significant. The farmer's social capital variables such as membership in farmer organisations and seed exchange experience were positive and statistically significant. The results indicate the social capital of the farmers in accessing greater information. The information and communication variable (ICT), ownership of cell phone, had a positive and statistically significant impact on accessing effective information.

The dummy variable, NFSM, to capture the differences in information available in NFSM districts and non-NFSM districts was positive but statistically insignificant. Therefore, the NFSM did not have a statistically significant impact in ensuring higher effective information about SRI to farmers.

The results for intensity of adoption of SRI and the depth of SRI adoption further strengthened the importance of social capital variable—membership in farmer organisation—in the adoption of SRI. The results from both the models showed that younger farmers were more enthusiastic in adopting SRI rather than older farmers. The negative and significant relationship between the age of the head of the household and the adoption of technology is observed in some studies (Adesina and Zinnah 1993; Teklewold et al. 2013; Manda et al. 2015). However, there are few studies that observed a positive relationship between the two as well (Langyintuo and Mungoma 2008; Meshram et al. 2012; Kassie et al. 2013). As hypothesised, availability of labour and effective information was significant in influencing the intensity of adoption. The irrigation facility did not come out to be significant in the adoption of SRI in terms of acres indicating the importance of irrigation and irrigation management only in certain components of SRI, i.e. alternate wetting and drying.

More number of variables were significant in affecting the depth of adoption model than intensity of adoption model. This is not surprising as many factors (e.g. irrigation facility as mentioned earlier and the cost of labourers) will appear to be more relevant when farmers adopt greater number of SRI practices. The yield realisation under full adoption of SRI is significantly higher than partial adoption (Palanisami et al. 2013). Irrigation facility was significant and positive in affecting the depth of adoption. Assets owned were also significant and positive. Farm assets such as mechanical hand weeders are more relevant only when farmers decide to adopt the SRI component of mechanical hand weeders to aerate the soil. Similar to intensity model, availability of labourers and access to effective information came out to be positive and significant in the adoption model as well.

Wage rates for labourers were crucial in determining the depth of adoption of SRI. Wage rates of women labourers were negatively related to adoption, whereas wage rate for male labourers was positively related to adoption. This is perhaps due to the fact that the shift away from manual weeding to mechanical weeding creates more demand for male labourers. So, the skill-intensive nature of mechanical weeding leads to higher demand for male labourers and thus higher wages. It points to the possibility of a gender-biased technical change.

Although the NFSM dummy variable had a positive sign in all the models, the variable was significant neither in the information models nor in the adoption models. The results pointed out the lacuna in government intervention in disseminating the awareness and the merits of SRI among farmers. Therefore, the study points out the urgent need for scaling up of the programmes to encourage more and more farmers to adopt SRI. The study also shows the importance of irrigation facility; so, there is a need for major investments in upgrading of irrigation scheme. The findings also revealed that the adoption of SRI is at the expense of women employment. Therefore, gender implications of SRI adoption need to be evaluated as a matter of some priority. The capacity building and targeted training are highly required not just for farmer groups but also for women agricultural labourers to upgrade their skills and expertise in using some of the farm machineries, especially the mechanical weeders.

Appendix

See Table 7.7.

Table 7.7 Descriptive statistics for SRI information from agricultural departments and NGOs

State	Mean	No. of observations	Districts	Mean	No. of observations
Karnataka	0.76 (0.42)	82	Chikmagalur	0.86 (0.35)	42
			Hassan	0.68 (0.47)	40
Madhya Pradesh	0.74 (0.43)	124	Shahdol	0.8 (0.40)	60
			Sidhi	0.68 (0.47)	64
Orissa	0.77 (0.42)	180	Keonjhar	0.74 (0.43)	98
			Mayurbhanj	0.80 (0.39)	82

Note Standard deviation is given in parentheses
Source Survey data

Chapter 8
Adoption of Multiple SRI Practices

8.1 Introduction

Despite an increase in the number of studies that describe SRI adoption and its benefits, our understanding of what drives a farmer to adopt different SRI components remains limited. A better understanding of factors that influence farmer's adoption of various SRI practices is, therefore, important for designing policies that could stimulate their adoption and thereby rice productivity and farm income. Farmers often choose only parts of a package or apply combination of practices only on small parts of their cultivated area (Smale et al. 1995). Also technology adoption decisions are interdependent, and combination of practices may influence each other (Teklewold et al. 2013). Attempting univariate modelling would exclude useful economic information about interdependent and double-hurdle adoption decisions (Dorfman 1996). Therefore, an analysis of factors influencing adoption decisions without controlling for technology interdependence and double-hurdle adoption might lead to inconsistent parameter estimates (Teklewold et al. 2013).

There have been attempts to model the interrelationship in the adoption of multiple agricultural technologies, with one of the pioneering attempts being made by Feder (1982). In recent years, more studies have looked at the joint estimation of multiple agricultural technologies (Teklewold et al. 2013; Manda et al. 2015). The study by Teklewold et al. (2013) applied a multivariate and ordered probit model to the household adoption of various sustainable agricultural practices in rural Ethiopia.

Not much attempts have been made to analyse the factors influencing the adoption of various practices/components of SRI using a multivariate framework, which is especially true for India. This chapter contributes to the growing adoption literature on SRI by making a detailed analysis of the factors influencing the adoption of various practices/components of SRI.

P. Varma, *Rice Productivity and Food Security in India*,
DOI 10.1007/978-981-10-3692-7_8

The remainder of the chapter is organised as follows. Section 8.2 provides conceptual and econometric framework for the analysis along with variable description and expected relationships. Section 8.3 presents the descriptive statistics along with the discussion of explanatory variables used in this study. The analytical results are discussed in Sect. 8.4. Concluding observations and policy implications are presented in Sect. 8.5.

8.2 Conceptual and Econometric Framework

Similar to existing studies on agricultural technology adoption behaviour, this study makes use of a random utility theory to explain adoption where the utility of a farm household is specified as a linear function of the household and farm-specific characteristics, institutional factors, attributes of technology as well as a stochastic component (Marenya and Barrett 2007). Farmers will adopt a practice or a combination of a practice that can provide maximum utility to them.

The probability of choosing a specific practice or a combination of practices is equal to the probability that the utility of that particular alternative is greater than or equal to the utilities of all other alternatives in the choice set. In order to maximise the utility U_{ij}, an ith farmer will compare alternative practices and combinations. Accordingly, an ith farmer will choose a practice j, over any alternative practice, k, if $U_{ij} > U_{ik}$, $k \neq j$.

As done in the case of the analysis adoption of depth and intensity of SRI adoption in Chap. 7, this chapter makes use of CMP estimator developed by Roodman (2011) (see Chap. 7 for more details). Farmers' choice of different interrelated SRI practices is modelled using a multivariate probit model (MVP), and the factors influencing the extent of combinations of SRI practices adopted are modelled using an ordered probit model.

8.2.1 Multivariate Probit Model

Decision to adopt different practices or components is inherently a multivariate decision. In single equation statistical models, farmers' inability to access one set of services does not alter the likelihood of them accessing another set of services. However, the multivariate probit model (MVP) simultaneously models the influence of the set of explanatory variables on each of the different technology practices, while allowing for the potential correlation between unobserved disturbances as well as the relationship between the access to different practices (Teklewold et al. 2013). The MVP recognises the correlation in the error terms of adoption equations and estimates a set of binary probit models simultaneously. The possibility for correlation is due to the fact that the same unobserved characteristics of farmers could influence the adoption of different SRI practices (Kassie et al. 2015). Failure to capture the interdependence of adoption of different SRI practices might lead to inconsistent parameter estimates (Kassie et al. 2015).

The farmer decides to adopt a k_{th} SRI practice with a latent (unobservable) dependent variable (Y_{ik}) as a function of a set of observable household, farm, institutional and other relevant factors and multivariate normally distributed error terms (ϵ_i) (Teklewold et al. 2013; Kassie et al. 2015). The same can be expressed as:

$$Y_{ik} = X_{ik}\beta_k + \varepsilon_{\text{ik}}(k = 1\ldots K) \tag{8.1}$$

where Y_{ik} denotes the latent dependent variable which can be represented by the level of expected benefit and/or utility derived from adoption. X represents a set of household, farm and institutional factors, and β is the parameter that needs to be estimated. $\in_{ik,}$ k = 1 … K are the multivariate normally distributed error terms. 1 to K practices are transplanting of young seedlings (t), shallow planting of seedlings (p), single seedling at wider spacing (s), use of organics (o), weeding by mechanical weeder (e) and efficient water management (m).

The second system of equations describing the observable binary outcome equation variables for each of the SRI practices choice of households is given as:

$$Y_{ik} = \begin{cases} 1 & \text{if } Y_{ik} > 0 \\ 0 & \text{otherwise} \end{cases} \tag{8.2}$$

In line with Teklewold et al. (2013) and Kassie et al. (2015), we describe a multivariate model, with the error terms jointly following a multivariate normal distribution (MVN) with zero conditional mean and variance normalised to unity for identification of parameters. This can be expressed as:

(ut, up, us, uo, ue, um)˜.MVN (0, Ω) and the symmetric variance-covariance matrix Ω is given by:

$$\Omega = \begin{bmatrix} 1 & ptp & pts & pto & pte & ptm \\ ppt & 1 & pps & ppo & ppe & ppm \\ pst & psp & 1 & pso & pse & psm \\ pot & pop & pos & 1 & poe & pom \\ pet & pep & pes & peo & 1 & pem \\ pmt & pmp & pms & pmo & pme & 1 \end{bmatrix}$$

where, p denotes the pair-wise correlation coefficient of the error terms with respect to any two SRI practice adoption equations. In the presence of the correlation of error terms, the off-diagonal elements in the variance-covariance matrix of adoption equations become nonzero, and Eq. (8.2) becomes an MVP model (Kassie et al. 2013).

8.2.2 *Ordered Probit Model*

The MVP model specified above only considers the probability of adoption of different practices of SRI, with no distinction made between, for example, those farmers who adopt one practice and those who adopt multiple practices in combination. Therefore, an ordered probit analyses the factors that influence the adoption of a combination of SRI practices (in terms of total number of practices adopted). Additionally, the variables that affect the adoption of a SRI practice may differently affect the intensity of adoption of practices (Teklewold et al. 2013). The probability of adopting first practice can differ from the probability of adopting a second practice or a third practice and so on. The farmers who adopt greater number of practices are definitely superior in terms of the intensity of adoption.

8.3 Descriptive Statistics: Description of Dependant and Explanatory Variables

Descriptive statistics for the six SRI practices as dependant variables and all explanatory variables are presented in Appendix Table 8.4; Fig. 8.1.

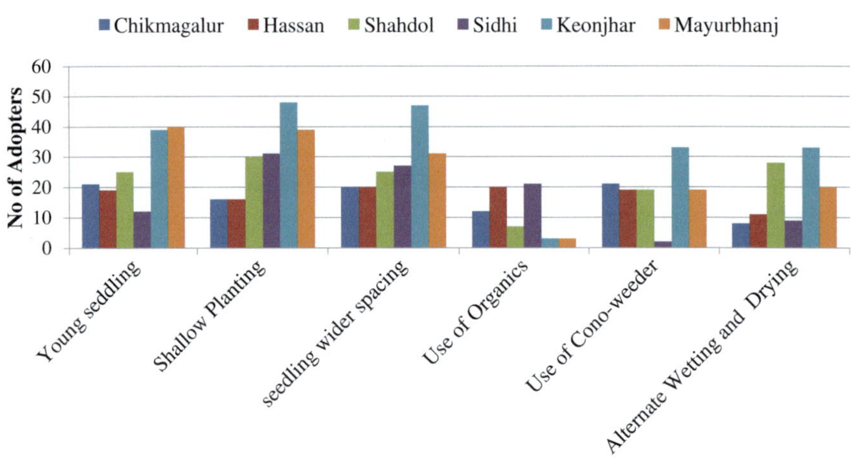

Fig. 8.1 Percentage of adoption of each practice/components of SRI among the adopted households. *Source* Survey data

8.3.1 The Dependent Variables

The number of adopters of young seedling was 156; shallow planting was 180; single seedling at wider space was 147; organics was 66; mechanical cono-weeders were 113; wetting–drying was 109. The highest adoption of SRI practices was observed in the Keonjhar district of Orissa. The lowest adoption of SRI practices was observed in the Hassan district of Karnataka.

8.3.2 Explanatory Variables

The model specification draws on the adoption literature (Feder et al. 1985; Adesina and Zinnah 1993; Moser and Barrett 2003; Langyintuo and Mungoma 2008; Pender and Gebremedhin 2008; Uaiene 2011; Meshram et al. 2012; Kassie et al. 2013; Teklewold et al. 2013; Ogada et al. 2014; Manda et al. 2015; Kassie et al. 2015).

We control for household heterogeneity by including variables such as age of the head of the household, gender of the head of the household, education level of the household and the family size.

Several of the explanatory variables used in the current model are also included in the earlier models that were discussed in Chap. 7. For example, the age and its implications on technology adoption are discussed in Chap. 7 (see Chap. 7 for more details). Similarly, a detailed discussion of variables such as gender of the household, size of the household, education of the household, labour, irrigation facility, terrain type and NFSM is already given in Chap. 7. So the sections below discuss the variables only briefly in order to avoid the overlap.

Age is included as a household variable that influences the adoption decisions in most of the literature (Feder et al. 1985; Uaiene 2011; Teklewold et al. 2013; Ogada et al. 2014; Manda et al. 2015; Kassiet et al. 2015). The average age of the head of the household in our sample ranges from fifty to fifty-one. Gender of the household head is also said to influence the adoption (Doss and Morris 2001; Langyintuo and Mungoma 2008; Pender and Gebremedhin 2008; Manda et al. 2015).

Education of the household is also important in affecting the technology adoption (Moser and Barrett 2003; Pender and Gebremedhin 2008; Langyintuo and Mungoma 2008; Haldar et al. 2012). The average number of an educated (education higher than 10th) person in the household was one. Similarly, the household size is used as a proxy to capture labour endowment (Pender and Gebremedhin 2008). The average numbers of active family labourers in our sample ranged from two to three, whereas the average number of household members was five.

As mentioned already in Chap. 7, agricultural extension services are an important channel for wider dissemination and adoption of technology (Langyintuo and Mungoma 2008; Kassiet et al. 2015), and this is especially true in the case of SRI as it is a knowledge-intensive innovation (Noltze et al. 2012). Similarly, market access also has huge bearing on transaction cost in accessing information and technology (Kassie et al. 2015). The average distance for households to main market in our sample was ten to sixteen kilometres.

Similar to the earlier models, to proxy labour, our study makes use of the wage rates for male and female labourers. The descriptive statistics shows that there is a huge disparity between male and female wage rates. The average male wage rate varied between Rs. 147 per day and Rs. 208 per day, whereas the female wage rate varied between Rs. 122 per day and Rs. 170 per day.

Importance of irrigation and irrigation management (Tsujimoto et al. 2009; Noltze et al. 2012; Uphoff 2012), terrain type (Moser and Barrett 2003), etc. are also discussed in greater detail in the Chap. 7. Similar to all our models, we also include dummy variables of NFSM to capture the impact of NFSM in promoting the adoption of SRI. The discussion below is on some of the explanatory variables that are not included in the models in Chap. 7. So we discuss them in a little more detail.

Although farm size is included in the earlier models, the present chapter defines farm size as marginal, small, semi-medium, medium and large unlike the earlier models where the total size of the landholding was used to capture the farm size (see Chap. 7). Majority of the farmers in our sample were marginal and small farmers with less than one hectare of land. In addition to the farm size, we also include number of farm assets owned and rented to capture the household wealth (Kassie et al. 2015). As per existing studies, household asset endowments play crucial role in adoption decisions (Kassie et al. 2015). Several studies noted a positive relationship between the farm assets and adoption (Langyintuo and Mungoma 2008; Kassie et al. 2015).

We also include a variable called number of years in rice cultivation to capture relationship between experience in agriculture and adoption. The average years of experience in our sample ranged between twenty and twenty-four years.

Another important human capital which is relevant in influencing the adoption and the extent of adoption is number of active family labourers (Langyintuo and Mungoma 2008; Noltze et al. 2012). Adoption of a new technology can be less attractive to those who do not have sufficient family labourers (Langyintuo and Mungoma 2008). Membership in input supply cooperatives is another important factor that has been highlighted in the literature in affecting the adoption decisions (Barrett et al. 2004; Matuschke and Qaim 2009; Khonje et al. 2015).

We give a dummy variable equal to 1 for those who have access to extension services. Although SRI is not specific to any particular variety of rice, seed exchange experience can also play a significant role in adoption decisions in general (Langyintuo and Mungoma 2008; Mazvimavi and Twomlow 2009; Shiferaw et al. 2015). Similar to extension services, we give a dummy variable equal to 1 for those who have experience in seed exchange.

Aversion to risk by farmers has also been highlighted by some studies as the reason for poor adoption (Johnson and Vijayaraghavan 2011). To capture this, we include a variable, fear of poor yield, in our model. Also, literature shows that soil type, source of irrigation, etc. are important in influencing the level of adoption (Palanisami et al. 2013). In our sample, around 258 farmers (around 67%) had the fear of poor yield. We give a dummy variable equal to 1 for those who have fear of poor yield. Proper soil management and water management are important not only in adoption but also in realising full potential of yield. Although SRI requires less water, moisture saturated but not flooded conditions require proper water management (Noltze et al. 2013). A detailed description of variables is given in Appendix Table 8.3.

8.4 Results and Discussion

The results for MVP are presented in Table 8.1. The model fits the data reasonably well; the hypothesis that all regression coefficients in each equation are jointly equal to zero is rejected. As expected, the likelihood ratio test [chi2(114) = 397.78, $p = 0.000$)] of the null hypothesis that the covariance of the error terms across equations is not correlated is also rejected.

The results showed that the assets owned by farm households is one of the most important factors determining the adoption of all SRI practices. The assets owned had a positive and significant impact on all SRI practices. Existing literature on technology adoption also highlights the significance of household assets (Langyintuo and Mungoma 2008; Kassie et al. 2015). Another variable that came out to be statistically significant in affecting the adoption of all practices was the wage rates for female labourers. In the case of almost all practices, the female wage rate had a negative and significant impact on adoption. The male wage rate was also significant in the adoption of two out of six practices. But the relationship was positive. The weeding operations under conventional rice cultivation have been traditionally done by women. However, as a result of SRI adoption, rice farmers hire more and more of male labourers for mechanical weeding (Senthikumar et al. 2008). Therefore, the shift from manual weeding to mechanical weeding might have resulted in greater demands for male labourers and thus a positive relationship between male wage rates and SRI adoption. It may be noted that the male wage rate was positive and significant in the adoption of cono-weeder for weeding under the SRI practice. Similarly, a decline in the demand for female labourers could be the reason for a negative relationship between female wage rates and SRI adoption.

Access to extension services and availability of proper irrigation facility also had significant impact in affecting the adoption decisions of most of the practices. Existing studies on technology adoption also highlight the importance of extension services in wider dissemination and adoption of technology (Langyintuo and Mungoma 2008; Devi and Ponnarasi 2009; Noltze et al. 2012; Kassiet et al. 2015). Similarly, studies have also highlighted the importance of irrigation in the adoption of SRI. As noted by Noltze et al. (2013), SRI necessitates proper water management. Other studies have also highlighted the importance of irrigation and irrigation management in affecting adoption of SRI (Tsujimoto et al. 2009; Noltze et al. 2012; Uphoff 2012). Nonetheless, the relationship between the availability of irrigation facility and the adoption of organic matter to enrich soils was negative and

Table 8.1 Result for multivariate probit model

A. Young seedling	Coefficient	Z-statistic	Marginal effects
Age of the HOH	−0.00 (0.01)	−0.09	**−0.00**
No. of family members	0.03 (0.05)	0.68	0.01
Active family labourers	0.08 (0.07)	1.07	0.02
No. of years in agriculture	−0.01 (0.01)**	−2.11	−0.00**
Farm size	−0.18 (0.10)*	−1.76	−0.06*
Contact extension service	0.31 (0.18)*	1.66	0.10*
Membership in input supply cooperatives	0.41 (0.19)**	2.15	0.13**
Distance from main market	0.01 (0.01)	1.28	0.00
Seed exchange experience	0.14 (0.14)	0.98	0.05
Wage rate for male	0.01 (0.00)**	2.34	0.00**
Wage rate for female	−0.01 (0.00) ***	−3.02	−0.00***
Fear of poor yield	−0.38 (0.16)**	−2.39	−0.12**
Education	0.04 (0.07)	0.56	0.01
Irrigation facility	0.29 (0.15)**	1.97	0.09**
Gender of HOH	−0.29 (0.26)	−1.10	−0.09
Assets owned	0.11 (0.03)***	3.34	0.03***
Assets rented	0.28 (0.10)***	2.88	0.09***
Terrain type	0.02 (0.21)	0.11	0.01
NFSM	−0.00 (0.16)	−0.01	−0.00
Constant	−0.28 (0.52)	−0.53	
B. Shallow planting			
Age of the HOH	0.00 (0.01)	0.26	0.00
No. of family members	0.03 (0.05)	0.69	0.01
Active family labourers	0.09 (0.07)	1.34	0.03
No. of years in agriculture	−0.01 (0.01)**	−2.11	−0.01**
Farm size	−0.03 (0.10)	−0.30	−0.01
Contact extension service	0.34 (0.18)**	1.93	0.12**
Membership in input supply cooperatives	0.22 (0.18)	1.24	0.08
Distance from main market	0.01 (0.00)	1.26	0.00
Seed exchange experience	0.11 (0.14)	0.75	0.04
Wage rate for male	0.00 (0.00)	0.64	0.00
Wage rate for female	−0.01 (0.00)**	−2.45	−0.00**
Fear of poor yield	−0.19 (0.15)	−1.24	−0.06
Education	−0.03 (0.08)	−0.38	−0.01
Irrigation facility	0.32 (0.14)**	2.22	0.11**
Gender of HOH	−0.14 (0.25)	−0.57	−0.05
Assets owned	0.13 (0.03)***	3.81	0.04***

(continued)

Table 8.1 (continued)

A. Young seedling	Coefficient	Z-statistic	Marginal effects
Assets rented	0.19 (0.10)*	1.84	0.06*
Terrain type	−0.02 (0.21)	−0.11	−0.01
NFSM	−0.09 (0.15)	−0.58	−0.03
Constant	0.01 (0.51)	0.02	
C. Single seedling wider space			
Age of the HOH	−0.00 (0.01)	−0.14	−0.00
No. of family members	0.06 (0.05)	1.27	0.02
Active family labourers	0.01 (0.07)	0.11	0.00
No. of years in agriculture	0.00 (0.01)	0.05	0.00
Farm size	0.02 (0.10)	0.23	0.01
Contact extension service	0.25 (0.19)	1.30	0.07
Membership in input supply cooperatives	0.08 (0.19)	0.41	0.02
Distance from main market	0.01 (0.01)***	2.73	0.00***
Seed exchange experience	0.22 (0.15)	1.48	0.06
Wage rate for male	0.00 (0.00)	0.25	0.00
Wage rate for female	−0.01 (0.00)**	−2.43	−0.00**
Fear of poor yield	−0.08 (0.17)	−0.50	−0.02
Education	0.05 (0.08)	0.61	0.01
Irrigation facility	0.14 (0.15)	0.96	0.04
Gender of HOH	−0.07 (0.27)	−0.27	−0.02
Assets owned	0.10 (0.03)***	2.78	0.03***
Assets rented	−0.04 (0.12)	−0.31	−0.01
Terrain type	0.12 (0.22)	0.56	0.04
NFSM	0.07 (0.15)	0.48	0.02
Constant	−0.60 (0.52)	−1.15	
D. Use of organics			
Age of the HOH	−0.01 (0.01)	−082	−0.00
No. of family members	0.02 (0.06)	0.39	0.00
Active family labourers	0.13 (0.09)	1.48	0.02
No. of years in agriculture	−0.02 (0.01)**	−2.56	−0.00***
Farm size	0.33 (0.12)***	2.71	0.06***
Contact extension service	0.62 (0.24)***	2.62	0.12***
Membership in input supply cooperatives	0.08 (0.24)	0.31	0.01
Distance from main market	0.02 (0.01)***	3.34	0.00
Seed exchange experience	−0.12 (0.18)	−0.69	−0.02
Wage rate for male	0.00 (0.00)	0.41	0.00
Wage rate for female	0.01 (0.00)	1.62	0.00
Fear of poor yield	−0.26 (0.19)	−1.32	−0.05

(continued)

Table 8.1 (continued)

A. Young seedling	Coefficient	Z-statistic	Marginal effects
Education	−0.18 (0.09)**	−1.97	−0.03**
Irrigation facility	−0.39 (0.18)**	−2.21	−0.07**
Gender of HOH	−0.01 (0.30)	−0.02	−0.00
Assets owned	0.20 (0.04)***	5.08	0.04***
Assets rented	0.15 (0.12)	1.23	0.03
Terrain type	−0.06 (0.25)	−0.23	−0.01
NFSM	−0.95 (0.20) ***	−4.64	−0.18***
Constant	−1.78 (0.58) ***	−3.09	
E. Use of cono-weeder			
Age of the HOH	−0.01 (0.01)	−0.81	−0.01
No. of family members	0.03 (0.05)	0.64	0.01
Active family labourers	0.06 (0.07)	0.75	0.01
No. of years in agriculture	−0.01 (0.01)	−0.98	−0.00
Farm size	−0.05 (0.11)	−0.43	−0.01
Contact extension service	0.61 (0.20)***	3.09	0.16***
Membership in input supply cooperatives	0.31 (0.19)	1.61	0.08
Distance from main market	0.01 (0.01)*	1.73	0.00*
Seed exchange experience	0.15 (0.15)	0.98	0.04
Wage rate for male	0.01 (0.00)***	4.05	0.00***
Wage rate for female	−0.01 (0.00) ***	−3.17	−0.00***
Fear of poor yield	−0.24 (0.17)	−1.43	−0.06
Education	0.01 (0.08)	0.09	0.00
Irrigation facility	0.32 (0.16)**	2.06	0.09**
Gender of HOH	−0.56 (0.28)**	−1.97	−0.15**
Assets owned	0.13 (0.03)***	4.06	0.03***
Assets rented	0.09 (0.11)	0.86	0.03
Terrain type	0.05 (0.21)	0.25	0.01
NFSM	0.26 (0.16)	1.58	0.07
Constant	−1.34 (0.54)**	−2.47	
F. Wetting and drying			
Age of the HOH	−0.01 (0.01)	−1.12	−0.00
No. of family members	0.03 (0.05)	0.52	0.01
Active family labourers	0.05 (0.08)	0.68	0.01
No. of years in agriculture	−0.00 (0.01)	−0.13	−0.00
Farm size	0.03 (0.11)	0.28	0.01
Contact extension service	0.39 (0.19)**	2.03	0.11**

(continued)

Table 8.1 (continued)

A. Young seedling	Coefficient	Z-statistic	Marginal effects
Membership in input supply cooperatives	0.34 (0.19)*	1.80	0.09*
Distance from main market	−0.01 (0.01)	−0.75	−0.00
Seed exchange experience	−0.12 (0.15)	−0.81	−0.03
Wage rate for male	0.00 (0.00)	1.12	0.00
Wage rate for female	−0.01 (0.00)**	−1.96	−0.00**
Fear of poor yield	−0.07 (0.17)	0.40	−0.02
Education	0.02 (0.08)	0.21	0.00
Irrigation facility	0.70 (0.16)***	4.48	0.19***
Gender of HOH	−0.29 (0.29)	−0.99	−0.08
Assets owned	0.11 (0.03)***	3.65	0.03***
Assets rented	0.13 (0.12)	1.04	0.04
Terrain type	−0.06 (0.23)	−0.26	−0.02
NFSM	0.08 (0.15)	0.53	0.02
Constant	−0.57 (0.55)	−1.04	
No of observations	386	LR chi2 (114)	397.78
Log likelihood	−755.19249	Prob > chi2	0.000

Notes ***, ** and * indicates statistical significance at 1, 5 and 10%, respectively. Standard errors are in parentheses

significant. This could be due to the fact that the adoption of organic matter is useful in improving the nutrients and water holding capacity of the soil and thus reduce the requirement of water.

Interestingly, the dummy variable for NFSM did not come out to be significant in almost all the models indicating the government's attempt to promote SRI practice through National Food Security Mission did not have desired impact in these districts. Additionally, the adoption of organic matter practice was higher in those districts where the SRI was not incorporated under NFSM.

The detailed results for SRI practices are given below.

8.4.1 Young Seedling

Membership in input supply cooperatives, wage rates, number of years in rice cultivation, assets owned and rented, irrigation facility, size of the farm, fear of poor yield, etc. had significant impact on adoption of young seedlings. Assets, access to extension services, membership in input supply cooperatives, better irrigation facility, etc. increase the likelihood of adoption. On the other hand, the experienced farmers or farmers who were in farming for many years were less likely to adopt young seedling for rice cultivation. The finding also showed that large farmers were less keen to adopt SRI, and this was evident from the negative and significant results for the relationship between size of the farm and the adoption of young seedlings. Similarly, fear of poor yield and high female wage rates also reduced the likelihood of adoption.

8.4.2 Shallow Planting

Household assets, access to extension services, female wage rates, number of years in rice cultivation and availability of irrigation were significant in the case of adoption of shallow planting. As seen in the case of young seedling, the adoption of shallow planting was also negatively related with the number of years farmers stayed in rice cultivation or experience in rice cultivation by farmers. The availability of extension services, household assets, availability of irrigation facility, etc. increased the likelihood of farmers adopting shallow planting. As noted in the case of young seedling, the higher wage rate of female labourers reduced the likelihood of farmers adopting shallow planting.

8.4.3 Single Seedling at a Wider Space

Household assets, female wage rates and the distance from main market came out to be significant in the adoption of single seedling at a wider space. Wage rate for female labourers reduced the likelihood of adopting single seedling at wider space. However, the results for distance showed that distance increased the likelihood of adoption. The results for distance from main market are counter-intuitive. But as mentioned earlier, majority of rice farmers were small or marginal farmers. Not only that the majority of rice farmers were semi-medium, small and marginal but also the adopters of SRI were semi-medium, small and marginal farmers. In addition to this, as observed from the discussions with farmers, majority of the rice farmers cultivated rice for self-consumption and retained very little for sale in the market. Therefore, adoption of SRI by medium, small and marginal farmers could be a strategy to improve household food security by enhancing rice yield. Also SRI does not require any particular variety of seed that needs to be purchased from the market. It does not require a costly innovation either. Therefore, it is possible that farmers who are remotely located from the market are more encouraged to shift to SRI than farmers who cultivate marketable crops and are close to main market

8.4.4 *Use of Organics*

Experience or number of years in rice cultivation, size of the farm, availability of extension services and irrigation facility, assets owned by the household, etc. came out to be significant in the case of adoption of organic matter to enrich the soil quality. As expected the size of the farm was positive and significant in the case of adoption of organic matter. This is in line with our hypothesis as the awareness of the benefits of organic farming in general is also high among the large farmers as compared to small farmers. In addition to this, the availability of extension services and assets owned by the farmers are also important. As mentioned earlier, the irrigation facility had a negative and significant relationship with the adoption of organic matter to enrich soils. This is, however, not surprising as the adoption of organic matter is expected to improve the soil nutrients and water holding capacity and therefore reduces the need for irrigation. In other words, the adoption of organic matter would be high in those areas where the irrigation facility is relatively lower. The dummy variable—NFSM—to capture the impact of the government's National Food Security Mission on the adoption of organic matter came out to be significant but negative. The negative and significant relationship between NFSM dummy and adoption of organics indicates that the adoption of organics was higher in non-NFSM districts such as Chikmagalur, Sidhi and Mayurbhanj.

8.4.5 *Use of Cono-Weeder*

Gender of the household head, contact with extension services, distance, wage rates, irrigation facility, assets, etc. were statistically significant in the adoption of cono-weeder. Extension services, male wage rates, distance, irrigation facility, assets, etc. increased the likelihood for adoption. On the other hand, the male head of the household, wage rate for female, etc. reduced the likelihood of adoption. The weeding operations under conventional rice cultivation have been traditionally done by women. Therefore, the availability of mechanical weeders might have encouraged the female-headed households to adopt more of mechanical weeders for weeding so that it allowed the women to be free to do something else.

8.4.6 *Water Management—Wetting and Drying*

Extension services, membership in input supply cooperatives, irrigation facility and assets increased the likelihood of households adopting water management. However, the female wage rates reduced the likelihood for adopting water management.

The results for ordered probit model are presented in Table 8.2. Although the magnitude of coefficients is slightly different, the same variables were significant in

Table 8.2 Ordered probit estimation for total practices

Variables	Coefficient	Marginal effects						
		$P(Y = 0/X)$	$P(Y = 1/X)$	$P(Y = 2/X)$	$P(Y = 3/X)$	$P(Y = 4/X)$	$P(Y = 5/X)$	$P(Y = 6/X)$
Age of the HOH	−0.01 (0.01)	0.00	1.21e−08	−4.96e−06	−0.00	−0.00	−0.00	−0.00
No. of family members	0.03 (0.04)	−0.01	−6.10e−08	0.00	0.00	0.00	0.00	0.00
Active family labourers	0.07 (0.06)	−0.03	−1.44e−07	0.00	0.01	0.01	0.01	0.00
No of years in agriculture	−0.01 (0.01)**	0.01**	3.14e−08	−0.00	−0.00**	−0.00**	−0.00**	−0.00**
Farm size	0.06 (0.09)	−0.02	−1.29e−07	0.00	0.00	0.01	0.01	0.00
Contact extension service	0.44 (0.16)***	−0.17***	−0.00	0.00	0.03**	0.06***	0.07***	0.02**
Membership in input supply cooperatives	0.29 (0.16)*	−0.12*	−0.00	0.00	0.02*	0.04*	0.05*	0.01
Distance from main market	0.01 (0.01)*	−0.00*	−2.17e−08	8.90e−06	0.00*	0.00*	0.00*	0.00*
Seed exchange experience	0.08 (0.13)	−0.03	3.92e−06	0.00	0.01	0.01	0.01	0.00
Wage rate for male	0.00 (0.00)*	−0.00*	−8.23e−09	3.38e−06	0.00	0.00	0.00*	0.00
Wage rate for female	−0.01 (0.00)**	0.00**	1.24e−08	−5.10e−06	−0.00*	−0.00**	−0.00**	−0.00*
Fear of poor yield	−0.18 (0.13)	0.07	0.00	−0.00	−0.01	−0.02	−0.03	−0.01
Education	0.03 (0.06)	−0.01	−7.40e−08	0.00	0.02	0.00	0.01	0.00
Irrigation facility	0.35 (0.13)***	−0.14***	1.13e−06	0.00	0.03**	0.04***	0.05***	.0.01**

(continued)

Table 8.2 (continued)

Variables	Coefficient	Marginal effects						
		$P(Y = 0/X)$	$P(Y = 1/X)$	$P(Y = 2/X)$	$P(Y = 3/X)$	$P(Y = 4/X)$	$P(Y = 5/X)$	$P(Y = 6/X)$
Gender of HOH	−0.23 (0.21)	0.09	0.00	0.00	−0.01	−0.03	−0.04	−0.01
Assets owned	0.11 (0.03)***	−0.05***	−2.49e−07	0.00	0.01***	0.01***	0.02***	0.00***
Assets rented	0.13 (0.07)*	−0.05*	−2.86e−07	0.00	0.01	0.02*	0.02*	0.01
Terrain type	−0.05 (0.17)	0.02	−2.82e−06	−0.00	−0.00	−0.01	−0.01	−0.00
NFSM	−0.06 (0.13)	0.02	5.22e−07	−0.00	−0.00	−0.01	−0.01	−0.00
Constant	0.12 (0.43)							
$\mu 2$	0.14 (0.43)							
$\mu 3$	0.20 (0.43)							
$\mu 4$	0.79 (0.43)							
$\mu 5$	1.34 (0.43)							
$\mu 6$	2.29 (0.45)							
No of observations	386	LR chi2 (19)	92.07					
Log Likelihood	−494.59242	Prob > chi2	0.000					

Notes ***, ** and * denote statistical significance at the 10, 5 and 1% levels, respectively. Standard errors in parentheses both models. The chi-squared statistic for the ordered probit model is 92.07 and is statistically significant, indicating that the joint test of all slope coefficients equal to zero is rejected. Results reveal that the number of SRI practices adopted increases with extension services, membership in input supply cooperatives, wage rates for male labourers, distance from main market, better irrigation facility, assets owned and rented, etc. On the other hand, the female wage rate and number of years or experience in rice cultivation reduced the likelihood of households adopting SRI practices. The results for NFSM were not only insignificant but also in most cases the sign came out to be negative.

8.5 Conclusion

We analysed the probability and level of adoption of multiple practices of SRI by rice farmers of selected districts in India using household-level data. We used MVP and ordered probit models to jointly analyse the adoption of multiple practices and the number of SRI practices adopted while recognising the interrelationship among them. Our approach extends the existing empirical studies by allowing for correlations across different practices of SRI. The results showed that various economic, institutional and access-related factors shape farmers' adoption of SRI practices.

The adoption of agricultural technologies in developing countries is mostly connected to farmers' economic ability to access new technologies. The importance of households assets in influencing the adoption decisions has been well documented in the literature (Awotide et al. 2012; Kassie et al. 2015). The present study found a significant and positive relationship between households' assets and adoption of SRI practices. Another very important revelation from our study is how certain fixed social bias and gender disparities are affecting the adoption. Despite considerable disparities in wage rates between male and female labourers, the present study showed that female wage rates reduced the likelihood of adoption of almost all practices. Interestingly, the male wage rate generally increased the likelihood of adoption. The weeding operations under conventional rice cultivation have been traditionally done by women. However, as a result of SRI adoption, rice farmers hire more and more of male labourers for mechanical weeding (Senthikumar et al. 2008). The shift from manual weeding to mechanical weeding results in greater demand for male labourers, and this could be the reason for a positive relationship between male wage rates and SRI adoption. The results also indicate the skill-intensive nature of SRI adoption and the gender implications of SRI adoption.

Information and extension services are very important driving forces for enhancing adoption of SRI. Knowledge and awareness are particularly relevant for innovations such as SRI where the adoption is more from sustainability perspective than immediate benefits (Noltze et al. 2012). Our results showed the importance of extension services in influencing adoption decisions. The insignificant impact of NFSM districts dummy on SRI adoption is an eloquent testimony to the fact that the objective of increased rice production by promoting SRI under the government's Food Security Mission did not yield the desired results. Additionally, farmers who are remotely located from the main market had higher likelihood of adopting SRI. This also indicates the possibility for cultivation of commercial crops by those farmers who have better proximity to market. Most farmers who were interviewed did not consider rice farming as a commercially viable venture; instead, the production was mainly for self-consumption and sale in the local markets.

The study also pointed out the importance of investment in infrastructure such as irrigation in promoting SRI. Although SRI requires less water as compared to traditional method, farmers require their own irrigation resources to adopt proper water management which is an essential component of SRI.

The significance of social capital and networks was also observed in our analysis. The membership in farmers' organisations such as input supply cooperatives increased the likelihood of SRI adoption. The significance of these factors indicates that policymakers should focus on establishing and strengthening local collective institutions. Local institutions can play a crucial role in providing farmers with timely information, inputs and technical assistance.

Appendix

See Tables 8.3 and 8.4.

Table 8.3 Variable definitions

Variable	Definition
Age of the HOH	Age of the head of the household
No. of family members	Number of the family members in the household including children
Active family labourers	Number of active family labourers
No. of years in agriculture	Number of years the household cultivating rice
Farm size	Marginal, small, semi-medium, medium and large size of the farm
Contact extension services	Availability of extension services to farmers, dummy variable = 1 if household receives any kind of extension services, = 0 otherwise
Membership in input supply cooperative(s)	Membership of any of the family member in input supply cooperatives, dummy variable = 1 if any of the family member has membership, = 0 otherwise
Distance from main market	Distance to the nearest main market in kilometres
Seed exchange experience	Experience of exchanging seeds among the farmers
Wage rate for male	Male wage rate for rice cultivation in rupees
Wage rate for female	Female wage rate for rice cultivation in rupees
Fear of poor yield	Fear of poor rice yield due to the adoption of SRI/any new method of rice cultivation. Dummy variable = 1 if the household has fear of poor yield due to SRI adoption, = 0 otherwise
Education	Number of family members who have education higher than 10th class
Irrigation facility	Dummy variable = 1 if the household had irrigation sources other than rainfall, = 0 otherwise
Gender of HoH	Gender of the head of the household, dummy variable = 1 if the household has a male head of the household
Assets owned	Total number of the type of farm assets (bullocks, tractors, machinery, etc.) owned by the household
Assets rented	Total number of the type of farm assets (bullocks, tractors, machinery, etc.) rented by the household
Terrain type	Dummy variable = 1 if the terrain is levelled, 0 otherwise
NFSM dummy	Dummy variable = 1 for those districts where SRI is incorporated within NFSM, = 0 otherwise

Table 8.4 Descriptive statistics of variables used in the analysis

Variables	All	Young seedling	Shallow planting	Single seedling at wider space	Use of organics	Use of cono-weeder	Wetting and drying
Age. of the HOH	50.16 (11.3)	50.42 (10.82)	50.12 (11.26)	50.67 (11.22)	48.89 (11.79)	50.69 (10.90)	50.24 (11.58)
No. of family members	5.10 (1.92)	5.20 (2.14)	5.24 (2.06)	5.45 (2.16)	4.85 (1.89)	5.17 (2.22)	5.29 (2.39)
Active family labourers	2.51 (1.26)	2.52 (1.34)	2.55 (1.30)	2.61 (1.40)	2.50 (1.18)	2.54 (1.39)	2.61 (1.43)
No. of years in Agriculture	24.7 (11.7)	23.04 (10.4)	23.18 (10.87)	25.08 (11.79)	20.33 (10.25)	23.65 (10.99)	24.76 (11.79)
Farm size	0.78 (0.84)	0.82 (0.83)	0.82 (0.86)	0.84 (0.90)	1.08 (0.95)	0.96 (0.91)	0.94 (0.93)
Contact extension service	0.47 (0.50)	0.60 (0.49)	0.58 (0.49)	0.58 (0.50)	0.56 (0.50)	0.70 (0.46)	0.70 (0.46)
Membership in input supply cooperative (s)	0.40 (0.49)	0.53 (0.50)	0.49 (0.50)	0.50 (0.50)	0.42 (0.50)	0.60 (0.49)	0.61 (0.49)
Distance from main market (in km)	11.48 (10.6)	12.31 (14.02)	12.32 (10.79)	13.27 (13.81)	16.92 (18.99)	12.81 (14.95)	10.39 (8.10)
Seed exchange experience	0.60 (0.49)	0.64 (0.48)	0.63 (0.48)	0.65 (0.48)	0.67 (0.48)	0.65 (0.48)	0.57 (0.50)
Wage rate for male	167.5 (75.5)	172.63 (74.15)	161 (65.87)	147.17 (56.01)	208.64 (83.59)	187.70 (81.42)	160.23 (68.67)
Wage rate for female	138.1 (59.8)	139.01 (53.43)	132.08 (46.10)	120.94 (39.44)	170.38 (50.75)	148.54 (59.21)	131.88 (46.29)
Fear poor yield	0.67 (0.47)	0.56 (0.50)	0.61 (0.49)	0.61 (0.49)	0.62 (0.49)	0.58 (0.50)	0.59 (0.49)
Education	0.79 (1.14)	1.02 (1.17)	0.86 (1.12)	0.88 (1.18)	1.06 (1.21)	1.12 (1.20)	0.96 (1.19)
Irrigation facility	0.50 (0.50)	0.62 (0.49)	0.62 (0.49)	0.62 (0.49)	0.48 (0.50)	0.65 (0.48)	0.77 (0.42)
Gender of HOH	0.90 (0.29)	0.89 (0.31)	0.91 (0.29)	0.91 (0.29)	0.95 (0.21)	0.88 (0.32)	0.90 (0.30)
Assets owned	1.4 (1.9)	2.01 (2.56)	1.96 (2.51)	1.90 (2.36)	2.30 (2.89)	2.26 (2.76)	2.08 (2.57)
Assets rented	0.58 (0.97)	0.81 (1.07)	0.64 (1.01)	0.42 (0.75)	1.02 (1.16)	0.88 (1.13)	0.65 (1.00)
Terrain type	0.29 (0.45)	0.35 (0.48)	0.30 (0.46)	0.28 (0.45)	0.47 (0.50)	0.41 (0.49)	0.35 (0.48)
NFSM	0.52 (0.50)	0.54 (0.50)	0.52 (0.50)	0.56 (0.50)	0.33 (0.48)	0.65 (0.48)	0.63 (0.48)
No. of observations	386	156	180	106	66	113	109

Note The numbers in parenthesis are standard deviations

Chapter 9
Impact of SRI Adoption on Rice Yield and Household Income

9.1 Introduction

Although SRI seems to offer a number of benefits, there is limited empirical evidence on the determinants of its adoption and/or their impacts on Indian farmers' welfare using a joint framework. There are hardly any studies analysing the factors that affect the decisions to adopt individual as well as the combinations of SRI principles and their impact using a multinomial selection framework. Therefore, this chapter makes an attempt to identify various factors that affect the decisions to adopt major principles of SRI—plant management, soil management and water management—as well as the combination of these principles and their impact on yield and household income.

The modelling technique adopted is a multinomial selection process where the expected benefits of SRI induce the adoption decisions (Manda et al. 2015). The study makes use of the multinomial endogenous treatment effects model (Deb and Trivedi 2006b) to account for selection bias due to both observed and unobserved heterogeneity and to assess the differential impacts of the adoption of different combinations of SRI. The multinomial endogenous treatment effects model allows the modelling of interdependency among the different components (Manda et al. 2015). Multinomial endogenous treatment effects model allows the distribution of the endogenous treatment (adoption of SRI) and outcomes (income and yield) to be specified using a latent factor structure, thereby allowing a distinction to be made between selection on observable and unobservable characteristics (Deb and Trivedi 2006b).

The remainder of this chapter is organised as follows: Section 9.2 describes conceptual framework and the multinomial endogenous treatment effects model. Section 9.3 discusses the variables and hypothesis used in the analysis, followed by Sect. 9.4 which presents the empirical results. The last section provides conclusion.

P. Varma, *Rice Productivity and Food Security in India*,
DOI 10.1007/978-981-10-3692-7_9

9.2 Conceptual and Econometric Framework

Adoption of agricultural technologies is not a simple yes or no decision. The technologies are usually introduced in packages that include several components. Farmers often choose only parts of a technology package or apply combination of practices only on small parts of their cultivated area (Smale et al. 1995). Therefore, an adoption decision is inherently multivariate. Attempting univariate modelling would exclude useful economic information about interdependent and simultaneous adoption decisions (Dorfman 1996). Farmers tend to adopt mix of components from an agricultural technology package to deal with a multitude of agricultural production constraints. In most cases, the different components may complement each other (Feder et al. 1985). There have been many attempts to model the interrelationship in the adoption of multiple agricultural technologies, of which Feder's (1982) was one of the pioneering attempts. In recent years, more studies have looked at the joint estimation of multiple agricultural technologies (Teklewold et al. 2013; Manda et al. 2015). The study by Teklewold et al. (2013) utilised multivariate and ordered probit models to the modelling of adoption decisions by farm households. Based on a random utility framework, a multinomial endogenous treatment effects model was applied by Manda et al. (2015).

In line with Manda et al. (2015), the present study makes use of a multinomial endogenous treatment effects model proposed by Deb and Trivedi (2006a, b). We consider the adoption of SRI as a choice over five combinations comprising three major principles such as soil management, plant management and water management.

Farmers will adopt a practice or a combination of practices that can provide maximum utility to them subject to various constraints. In order to maximise the utility U_{ij}, farmer will compare alternative practices and combinations. Accordingly, an *i*th farmer will choose a practice *j*, over any alternative practice, *k*, if $U_{ij} > U_{ik}$, $k \neq j$.

As farmers may endogenously self-select adoption or non-adoption, decisions are likely to be influenced systematically both by observed and unobservable characteristics that may be correlated with the outcomes of interest. To disentangle the pure effects of adoption and to effectively assess the adoption and impact of SRI, we adopt a multinomial endogenous treatment effects model proposed by Deb and Trivedi (2006b). This approach has the advantage of evaluating both individual and combined practices, while capturing the interactions between the choices of alternative practices (Wu and Babcock 1998; Mansur et al. 2008).

Model Specification

The multinomial endogenous treatment effects model consists of two stages. In the first stage of the model, a farmer will choose one of the three practices mentioned above. Following Deb and Trivedi (2006a, b), let U_{ij} denote the indirect utility associated with the *j*th SRI practice, $j = 0, 1, 2 \ldots, J$ for farmer *i*:

$$U_{ij}^{*} = Z_i'\alpha_j + \sum_{k=1}^{j} \delta_{jk} l_{ik} + n_{ij} \tag{9.1}$$

where Z_i denotes the vector of household, social, economic and institutional factors associated parameters α_j; η_{ij} are independently and identically distributed error terms. Also, U_{ij}^{*} includes a latent factor l_{ik} that incorporates unobserved characteristics common to farmer i's treatment choice and outcome variables. Outcome variables in our analysis are plant, soil and water management practices, whereas the unobserved characteristics that may have an impact on outcome variables are management and technical abilities of farmers in understanding new practices and other infrastructural and institutional constraints (Manda et al. 2015). The l_{ik} are assumed to be independent of n_{ij}. Following Deb and Trivedi (2006b), let $j = 0$ denotes the control group and $U_{I0}^{*} = 0$. The control group in our analysis is the non-adopters of SRI. Let d_j be the observable binary variables representing the choice of various practices and as a vector of $d_i = (d_{i1}, d_{i2}, \ldots ,d_{iJ})$. Similarly, let $l_i = (l_{i1}, l_{i2}, \ldots ,l_{iJ})$. Then, the probability of treatment can be represented as:

$$\Pr(d_i|z_i l_i) = g(z_i'\alpha_1 + \delta_1 l_{i1}, z_i'\alpha_2 + \delta_2 l_{i2} \ldots z_i'\alpha_j + \delta_j l_{ij}) \tag{9.2}$$

where g is a multinomial probability distribution and g is expected to have a mixed multinomial logit (MMNL) structure defined as:

$$\Pr(d_i|z_i l_i) = \frac{\exp(z'\alpha_j + \delta_j l_{ij})}{1 + \sum_{k=1}^{j} \exp(z_i'\alpha_k + \delta_k l_{ik})} \tag{9.3}$$

The analysis of the impact of adoption of SRI on household income and rice yield is undertaken in the second stage. The expected outcome equation can be defined as:

$$E(y_i|d_i x_i l_i) = x_i'\beta + \sum_{j=1}^{j} \Upsilon_j d_{ij} + \sum_{j=1}^{j} \lambda_j l_{ij} \tag{9.4}$$

where y_i represents the outcome variables—household income and rice yield for farmer i—whereas x_i is a set of exogenous variables with associated parameter vectors β, and γ_j denotes the treatment effects relative to the control group, i.e. non-adopters of SRI. The possible endogeneity in adoption decision of SRI would lead to inconsistent estimates Υ as we treat d_{ij} to be exogenous. The $E(y_i|d_i, x_i, l_i)$ is a function of each of the latent factors l_{ij}. This shows the outcome is affected by unobserved characteristics that also affect selection into treatment. When λ_j, the factor-loading parameter, is positive (negative), treatment and outcome are positively (negatively) correlated through unobserved characteristics. This implies that there is positive (negative) selection with γ and λ the associated parameter vectors, respectively. We assume a normal (Gaussian) distribution function, as in our case the outcome variables—household income and yield—are continuous variables. The model was estimated using maximum simulated likelihood (MSL) approach.[1]

Next step is of including valid instruments in the model. As per Deb and Trivedi (2006a), the parameters of the model are estimated even if the explanatory variables in the treatment equation are the same as the ones used in the outcome equation; the use of exclusion restrictions or instruments will provide more robust estimates. In our analysis, therefore, we include additional variables in the treatment equation that are not correlated with outcome variables. Although finding a valid instrument is empirically challenging, we use access to information regarding SRI from formal sources as our instrument variable.[2] The information regarding SRI can have an impact on the adoption decisions of SRI but is hardly expected to influence the outcomes such as farmers' income and rice yield. Several studies on adoption and impact of technology have utilised information as an instrument variable (Di Falco and Bulte 2011; Di Falco and Veronesi 2013; Manda et al. 2015). We also ran a test to see the validity of instrument, and the results are presented in Appendix Table 9.4. The instrument variable—information—that we selected has an impact on the adoption decision by farmers in most cases but do not have any impact on the outcome variables of the non-adopters.

9.3 Description of Variables and Hypothesis

The adoption models include several explanatory variables based on economic theory and empirical literature on adoption. The definition of the variables used in the current econometric analysis is presented in Appendix Table 9.4. The details of the explanatory variables are as follows:

The factors that are likely to affect adoption decisions include household and farm characteristics (age of the household head, education, household size, etc.); social capital (membership in farmer association); availability of extension services; crop stresses (fear of poor yield); plot characteristics (type of terrain); and location characteristics (whether the district belongs to NFSM). Therefore, several studies have included household characteristics such as age of the head of the household and the size of the household, education as important factors influencing the adoption decision by farmers (Feder et al. 1985; Uaiene 2011; Teklewold et al. 2013; Ogada et al. 2014). A detailed discussion of these variables and its expected impact is given in Chaps. 7 and 8.

---

[1]The model was estimated using the Stata command mtreatreg, which is an extension of the treatreg Stata command to a multinomial approach by Deb (2009) and 200 simulation draws were used.

[2]Here we define information regarding SRI as information about SRI from formal sources such as government departments and NGOs. We omit those farmers who had information about SRI through informal sources such as neighbours and other farmers.

Similar to the models in Chap. 8, the present chapter has included the variable—experience in agriculture—to capture the relationship between the experience in agriculture and SRI adoption. We include a variable called no. of years in agriculture in our analysis. The average years of experience in our sample ranged between twenty to twenty-four years. Similarly, the number of active family labourers is considered to be an important human capital in influencing the adoption and the extent of adoption (Langyintuo and Mungoma 2008; Noltze et al. 2012) (see Chap. 8).

As discussed in the preceding chapter, the farm assets owned by the farmer are also important in deciding the adoption (Langyintuo and Mungoma 2008; Teklewold et al. 2013). The households' ability to cope up with production and price risk increases with an increase in wealth or stock of productive assets. Similarly, access to off-farm activities and income is expected to have a positive impact on adoption decisions (Davis et al. 2009).

Farmer's awareness of the benefits of a new technology stems from the fact that they have access to information and extension services. Since SRI is a knowledge-based innovation, extension services play an even greater role in wider adoption (Noltze et al. 2012). In order to make the information available to farmers in an effective manner, regular visits to the field and guidance are very important (Langyintuo and Mungoma 2008). There are studies that point out the importance of access to information in determining the adoption decision even for a farmer with positive demand for adoption (Langyintuo and Mungoma 2008; Mazvimavi and Twomlow 2009; Shiferaw et al. 2015). Additionally, the quality of information is also very important. In our study, we noticed that most farmers knew about SRI and only 38 farmers in our sample (hardly 10%) did not hear about SRI. Around 139 farmers received information regarding SRI directly from government departments and 155 farmers received information regarding SRI directly from NGOs. Around 54 farmers knew about SRI from other sources mainly neighbours. In our study, we give dummy variable equal to one if the household received information regarding SRI from agricultural departments and NGOs and zero otherwise. The descriptive statistics also show that the information regarding SRI practices from agricultural departments and NGOs was the lowest among the non-adopters and the average information received from adopters of all the practices from agricultural departments and NGOs was quite high.

There are studies that point out the importance of availability of seed in determining the adoption decision (Langyintuo and Mungoma 2008; Mazvimavi and Twomlow 2009; Shiferaw et al. 2015). We use the number of rice varieties known by the household as a proxy to capture the impact of availability of seed on the adoption of SRI.

Similar to the earlier models, aversion to risk by farmers is also included in the present analysis. To capture this, we include a variable, fear of poor yield, in our model. In addition, the literature shows that soil type, source of irrigation, etc., are important in influencing the level of adoption (Palanisami et al. 2013). Proper soil and water management are important not only in adoption but also in realising full potential of yield. Although SRI requires less water, moisture saturated but not flooded conditions require proper water management (Noltze et al. 2013). Therefore, studies have highlighted the importance of irrigation and irrigation management as important factors in deciding adoption of SRI (Tsujimoto et al. 2009; Noltze et al. 2012; Uphoff 2012; Takahashi 2013). Poor land terrain is cited as one of the most important deterrents of adoption of SRI (Bhatt 2015). Significant differences in adoption intensity between regions were noticed by some studies (Langyintuo and Mungoma 2008). So, we include terrain type in our model.

Membership in farmer organisation and distance from main market are also included in the present study. Interestingly, average distances to main market for adopters of some of the practices or combinations of SRI practices (e.g. plant management and plant management plus soil management) were more than the non-adopters (see Table 9.2).

Finally similar to earlier models, in order to capture the impact of NFSM in promoting the adoption of SRI, we incorporate NFSM variable in the study.

9.4 Estimation Results and Discussion

9.4.1 Descriptive Statistics

Descriptive statistics of the explanatory variables that are hypothesised to influence adoption and outcome variables—yield and income—are presented in Table 9.1. As mentioned earlier, here we focus on SRI adoption as a choice over five alternatives involving three major SRI principles. The first group is non-adopters which had 193 households. The total number of adopters of SRI is also 193. Among the adopters, around 89 households (46%) have adopted all three components of SRI. Plant management and soil management are adopted by around 47 farmers (24%). Similarly, 38 households (20%) have adopted plant management alone and 19 households (10%) have adopted plant management and water management alone.

The descriptive statistics show that the level of education (average number of family members with education higher than tenth standard) was generally higher among the adopters of SRI than among non-adopters. Similarly, the average number of farm assets owned by the non-adopters of SRI was also less as compared to the adopters. Membership in farmer organisations was generally lower for non-adopters and for those practicing only plant management. Fear of poor yield was highest among the non-adopters than adopters. As expected, the access to extension services was the lowest among the non-adopters. Access to off-farm activities was generally better among the non-adopters of SRI. Similarly, the availability of irrigation facility was relatively less among the non-adopters of SRI. Information regarding SRI was also one of the lowest among the non-adopters of SRI (see Table 9.1).

Table 9.1 Descriptive statistics for variables used in the model

Variable	Non-adopters of SRI	Plant management	Plant and water management	Plant and soil management	Plant, soil and water management
No. of family members	4.97 (1.7)	5.61 (1.59)	5.37 (2.29)	4.77 (1.42)	5.28 (2.43)
Active family labourers	2.4 (1.2)	2.63 (1.26)	2.68 (1.34)	2.28 (1.02)	2.61 (1.46)
Education	0.68 (1.13)	0.63 (1.13)	1.05 (1.22)	0.94 (1.05)	0.96 (1.19)
Size of the landholding	3.0 (2.7)	2.31 (1.64)	4.15 (5.22)	3.31 (1.87)	3.87 (3.24)
Assets owned	0.97 (0.71)	1.39 (1.82)	2.11 (2.60)	2.32 (3.01)	2.09 (2.59)
Irrigation facility	0.40 (0.49)	0.47 (0.51)	0.63 (0.50)	0.30 (0.46)	0.81 (0.40)
Access to off-farm activity	0.38 (0.5)	0.26 (0.45)	0.21 (0.42)	0.36 (0.49)	0.22 (0.42)
Membership in farmer organisations	0.32 (0.47)	0.29 (0.46)	0.53 (0.51)	0.34 (0.48)	0.63 (0.49)
Improved varieties known	0.58 (0.79)	0.39 (0.55)	0.68 (1.00)	0.62 (0.71)	1.04 (0.93)
Terrain type	0.26 (0.44)	0.21 (0.47)	0.32 (0.48)	0.40 (0.50)	0.37 (0.51)
Fear of poor yield	0.73 (0.45)	0.68 (0.47)	0.68 (0.48)	0.62 (0.49)	0.56 (0.50)
Experience in agriculture	26.1 (12.34)	22.42 (9.43)	29.21 (14.36)	20.49 (9.61)	23.70 (11.03)
Access to extension service	0.36 (0.48)	0.39 (0.50)	0.52 (0.51)	0.43 (0.50)	0.73 (0.45)
Distance to main market	10 (6.3)	14.29 (10.73)	7.47 (0.40)	17.47 (21.6)	11.07 (8.6)
Have information about SRI	0.56 (0.50)	0.92 (0.27)	0.89 (0.31)	1 (0)	0.98 (.15)
NFSM dummy	0.52 (0.50)	0.29 (0.46)	0.74 (0.45)	0.45 (0.50)	0.61 (0.49)

Note Standard deviation is given in parentheses

9.4.2 Factors Influencing the Adoption of SRI

Table 9.2 presents parameter estimates of the mixed multinomial logit model which is equivalent to the first stage of our multinomial endogenous treatment effects model. The base category is non-adoption against which results are compared. The model fits the data very well with the Wald test, chi2 = 28.77; $P >$ chi2 = 0.02

Table 9.2 Mixed multinomial logit model estimates of adoption of SRI for yield and income (baseline category is non-adoption of SRI)

Variable	Plant management	Plant and water management	Plant and soil management	Plant, soil and water management
No. of family members	0.26 (0.14)*	0.10 (0.16)	0.03 (0.13)	0.09 (0.11)
Active family labourers	−0.04 (0.23)	−0.01 (0.21)	−0.12 (0.21)	0.10 (0.15)
Education	−0.20 (0.21)	0.20 (0.22)	−0.04 (0.19)	−0.05 (0.17)
Size of the landholding	−0.32 (0.17)**	0.03 (0.10)	−0.02 (0.08)	−0.05 (0.06)
Assets owned	0.45 (0.21)**	0.42 (0.18)**	0.61 (0.18) ***	0.51 (0.17)***
Irrigation facility	0.19 (0.47)	0.93 (0.55)*	−1.12 (0.48) **	1.74 (0.40)***
Access to off-farm activity	−0.87 (0.47)*	−0.76 (0.68)	−0.37 (0.46)	−0.56 (0.41)
Membership in farmer organisations	−0.15 (0.66)	0.48 (0.63)	−0.30 (0.62)	0.42 (0.46)
Improved varieties known	−0.45 (0.32)	−0.24 (0.31)	−0.22 (0.28)	0.41 (0.22)*
Terrain type	0.78 (0.61)	−0.42 (0.60)	0.92 (0.47)**	−0.03 (0.50)
Fear of poor yield	−0.20 (0.50)	0.06 (0.50)	−0.62 (0.47)	−0.77 (0.38)**
Experience in agriculture	−0.03 (0.02)	0.00 (0.03)	−0.07 (0.02) ***	−0.04 (0.01)***
Access to extension service	0.74 (0.70)	−0.76 (0.60)	−0.11 (0.73)	0.73 (0.56)
Ln distance to main market	0.62 (0.29)**	−0.71 (0.30) **	0.29 (0.40)	−0.21 (0.24)
NFSM dummy	−1.06 (0.47)**	0.51 (0.57)	−0.54 (0.46)	−0.46 (0.45)
Have information about SRI	−0.14 (0.62)	1.23 (0.69)*	1.38 (0.61)**	1.07 (0.51)**
Constant	−2.93 (1.09)***	−3.30 (1.46) **	−0.72 (1.20)	−2.32 (0.10)**
Wald chi2 = 28.77, $P >$ chi2 = 0.02				

Notes Sample size is 386 and 200 simulation draws were used. ***$P < 0.01$, **$P < 0.05$, *$P < 0.1$. Robust standard errors are given in parentheses

implying that the null hypothesis that all the regression coefficients are jointly equal to zero is rejected.

The results show that adoption of most packages increases with the number of farm assets owned by households, and it was significant and positive for all combinations. The relationship between access to information and SRI adoption was positive and significant in the case of three out of four SRI practices/combinations—

plant plus water management, plant management plus soil management and the plant management plus water management plus soil management. This is consistent with some studies on technology adoption in general (Adegbola and Gardebroek 2007; Manda et al. 2015) and in the specific context of SRI (Noltze et al. 2012). However, farm size generally had a negative impact and was significant in terms of the adoption of plant management indicating greater adoption of SRI by smaller farms. This contradicts the findings by Noltze et al. (2012) in his study on SRI in Timor Leste. Similarly, access to off-farm activity had a negative relationship with all practices/combinations of SRI and was significant in the case of the adoption of plant management. The negative relationship between access to off-farm activity and adoption of plant management is consistent with Pender and Gebremedhin (2008), Mathennge et al. (2015) and Manda et al. (2015). Better access to off-farm activities might divert the resources away from agriculture to off-farm activities and, therefore, result in less resource allocation for farm activities. The results show the adoption of SRI increases with household size but the relationship between household size and the adoption turned out to be statistically significant only in the case of the adoption of plant management. This is consistent with Noltze et al. (2012) and Teklewold et al. (2013).

Fear of poor yield was negative but statistically significant in the case of the adoption of all practices/combinations of SRI. Number of improved rice varieties known, etc., had a positive and significant impact on the full adoption (plant plus water plus soil management) of SRI. Availability of better irrigation facility had a statistically significant impact on the adoption of three practices/combinations of SRI—plant plus water management, plant plus soil management and plant plus soil plus water management. However, in the case of the adoption of the combination of plant plus soil management, there was a negative and statistically significant relationship with the availability of better irrigation facility. This is not counterintuitive. As obvious from the results, the lack of proper irrigation facility would be the reason for the non-adoption of water management while adopting SRI. These findings are consistent with Tsujimoto et al. (2009), Noltze et al. (2012), Uphoff (2012) and Takahashi (2013). Terrain type was statistically significant and positive in affecting the adoption of plant and soil management. This finding is consistent with Bhatt (2015) who highlights the poor land terrain as a deterrent of adoption of SRI.

Another interesting revelation from the analysis is that experienced farmers or those farmers who had been in rice cultivation for many years were not enthusiastic about adopting SRI. This is quite evident from the negative and statistically significant relationship between the variable—experience in agriculture—and the different combinations/practices of SRI. Although there are studies that argue that older farmers are more experienced and might have accumulated greater physical and social capital (Kassie et al. 2013), there is also an argument that older farmers are less amenable to change and, therefore, unwilling to change from old practices to new ones (Adesina and Zinnah 1993). Therefore, the findings from the current analysis are consistent with Adesina and Zinnah (1993) and Teklewold et al. (2013).

As far as the NFSM dummy is concerned, the results were negative in most cases and significant and negative in the case of the adoption of plant management. The results indicate that the government's attempt to promote food security by encouraging rice production through SRI did not yield the desired results.

9.4.3 Average Treatment Effects of Single and Different Combinations of SRI

Table 9.3 presents the estimates of the impact of SRI combinations on rice yields and household incomes. We estimated the impact of adoption on income and yield considering these outcome variables as both exogenous and endogenous. The results under the assumption of exogenous adoption showed that adopters had higher yield than non-adopters, and the results were statistically significant for all combinations of SRI except for plant management. However, the adoption of various practices/combinations of SRI did not have any impact on household income under the exogenous assumption. The inferences based on the exogenous assumption may be misleading as it ignores the effects of unobserved factors (Manda et al. 2015). Therefore, a multinomial endogenous treatment effects model is estimated to overcome this issue.

The average adoption effects after controlling for unobserved heterogeneity show similar results in terms of yield but not so in terms of income. All combinations of SRI had a positive and statistically significant impact on rice yield as compared to non-adopters, whereas the positive and significant impact of income was evident only in the adoption of two combinations/practices—plant management plus water management and plant management plus soil management. The adoption of plant management practice had a negative and statistically significant impact on income. Similarly, the full adoption of SRI also had a negative and statistically significant impact on income. The results are indicating that the adoption of some practices/combinations is costlier than others. In terms of other variables affecting the welfare outcomes, the exogenous model showed that number of active family labourers, education of the household, irrigation facility, access to off-farm activity, number of improved rice varieties known, terrain type, access to extension services, experience in agriculture, NFSM, etc., had a positive and statistically significant impact on rice yield. However, the fear of poor rice yield and the size of the farm had a negative impact on rice yield. In the case of household income, the exogenous model showed that number of active family labourers, education of the household, size of the farm, number of improved rice varieties known, type of terrain, etc., had a positive and statistically significant impact on household income. However, the number of family members (size of the household), farmer's fear of poor yield, aversion to risk, etc., had a negative and statistically significant impact on household income. As mentioned earlier, under the assumption of exogenous SRI adoption, the unobserved factors may be correlated with the outcome variables.

Table 9.3 Multinomial endogenous treatment effects model estimates of SRI impacts on household income and rice yield

SRI practice	Rice yield	Household income
Exogenous		
Plant management	0.03 (0.06)	−0.02 (0.03)
Plant management and water management	0.28 (0.06)***	−0.05 (0.06)
Plant management and soil management	0.18 (0.05)***	0.02 (0.04)
Plant management, water management and soil management	0.0.24 (0.04)***	−0.00 (0.03)
No. of family members	−0.01 (0.01)	−0.02 (0.01)***
Active family labourers	0.02 (0.01)**	0.02 (0.01)**
Education	0.06 (0.01)***	0.05 (0.01)***
Size of the landholding	−0.02 (0.01)***	0.04 (0.01)***
Assets owned	−0.00 (0.01)	0.00 (0.01)
Irrigation facility	0.06 (0.03)**	0.01 (0.02)
Access to off-farm activity	0.06 (0.03)*	0.03 (0.02)
Membership in input supply cooperative	0.03 (0.04)	0.02 (0.03)
Improved varieties known	0.05 (0.02)**	0.08 (0.02)***
Terrain type	0.12 (0.04)***	0.18 (0.03)***
Fear of poor yield	−0.10 (0.03)***	−0.09 (0.02)***
Experience in agriculture	0.00 (0.00)**	−0.00 (0.00)
Access to extension service	0.08 (0.04)*	0.00 (0.03)
Ln distance to main market	0.03 (0.02)*	0.00 (0.02)
NFSM dummy	0.15 (0.03)***	0.02 (0.02)
Constant	0.66 (0.09)***	−0.50 (0.06)***
Endogenous		
Plant management	0.20 (0.00)***	−0.15 (0.00)***
Plant management and water management	0.20 (0.00)***	0.02 (0.00)***
Plant management and soil management	0.19 (0.01)***	0.13 (0.00)***
Plant management, water management and soil management	0.28 (0.00)***	−0.08 (0.00)***
No. of family members	−0.02 (0.00)***	−0.02 (0.00)***
Active family labourers	0.03 (0.00)***	0.03 (0.00)***
Education	0.08 (0.00)***	0.04 (0.00)***
Size of the landholding	−0.02 (0.00)***	0.04 (0.00)***
Assets owned	0.00 (0.00)***	0.00 (0.00)***
Irrigation facility	0.02 (0.00)***	0.06 (0.00)***
Access to off-farm activity	0.01 (0.00)***	0.07 (0.00)***
Membership in farmer organisations	0.07 (0.00)***	0.02 (0.00)***
Improved varieties known	0.04 (0.00)***	0.08 (0.00)***
Terrain type	0.13 (0.00)***	0.20 (0.00)***

(continued)

Table 9.3 (continued)

SRI practice	Rice yield	Household income
Fear of poor yield	−0.13 (0.00)***	−0.11 (0.00)***
Experience in agriculture	0.00 (0.00)***	−0.00 (0.00)***
Access to extension services	0.06 (0.00)***	0.03 (0.00)***
Ln distance to main market	0.03 (0.00)***	−0.00 (0.00)***
NFSM dummy	0.16 (0.00)***	0.01 (0.00)***
Constant	0.72 (0.01)***	−0.49 (0.01)***
Selection terms (λ)		
Plant management	−0.19 (0.00)***	0.15 (0.00)***
Plant management and water management	0.15 (0.00)***	−0.08 (0.00)***
Plant management and soil management	−0.15 (0.00)***	−0.13 (0.00)***
Plant management, water management and soil management	−0.06 (0.00)***	0.12 (0.00)***

Notes The baseline is farm households that did not adopt any SRI. Sample size is 386 and 200 simulation draws were used. ***$P < 0.01$, **$P < 0.05$, *$P < 0.1$. Robust standard errors are given in parentheses

The results for endogenous model showed that number of active family labourers, education of the household, number of farm assets owned, irrigation facility, access to off-farm activity, membership in farmer organisations, number of improved rice varieties known, experience in agriculture, type of terrain, experience in agriculture, access to extension services, NFSM, etc., had a positive and statistically significant impact on rice yield. However, size of landholding and household size had a negative impact on rice yield. Similarly, number of active family members, size of the landholding, education of the household, number of farm assets owned, access to off-farm activity, access to extension services, membership in farmer organisation, type of terrain, irrigation facility, the number of improved rice varieties known, NFSM, etc., had a positive and statistically significant impact on household income. Distance from main market, fear of poor yield, household size, etc., had a statistically significant and negative impact on household income.

In addition, most of the factor loadings (k) show evidence of negative selection bias suggesting that unobserved factors that increase the likelihood of adopting different combinations/practices of SRI are associated with lower levels of welfare than those expected under random assignment to the different combinations/practices of SRI adoption status. Positive selection bias is also evident in the yield and income equation suggesting that unobserved variables increasing the likelihood of adopting residue retention are associated with higher levels of income and rice yield.

9.5 Conclusion

Although there are studies analysing the factors contributing to adoption of SRI and the welfare implications, this chapter contributes to the empirical literature in this area by examining the determinants and impacts of the adoption of five mutually exclusive combinations of SRI on yields and household incomes using a multinomial endogenous treatment effects model. As in earlier models, we find that the decision to adopt is a function of farm assets, irrigation facility, information about SRI from formal sources, experience in agriculture, number of improved rice varieties known, access to off-farm activity, etc. The number of farm assets came out to be significant in the adoption of all SRI practices/combinations. This was followed by the availability of information. The variable for information was positive and statistically significant in the adoption of three out of four SRI practices/combinations. The next important factor which came out to be significant in affecting the adoption was irrigation facility. The variable for irrigation facility came out to be significant in the adoption of three out of four SRI practices/combinations. In the case of plant and soil management practices, the irrigation facility had a negative and statistically significant impact. This indicates that the higher adoption of plant and soil management practices by households was due to the lack of proper irrigation facility.

Apart from the above factors, the other variables that affected the adoption of some of the practices/combinations of SRI were number of improved rice varieties known, household size, access to off-farm activity, fear of poor yield, distance to main market, type of terrain, size of the landholding and experience in agriculture. Number of improved rice varieties known, household size, type of terrain (terrain which is levelled), etc., increased the likelihood of household adopting different practices/combinations of SRI, whereas experience in agriculture or the number of years the farmer stayed in rice cultivation, access to off-farm activity, size of the landholding, etc., decreased the likelihood of adopting different practices/combinations of SRI. The dummy variable for NFSM was insignificant in most cases. In the case of the adoption of plant management, the NFSM was negative and statistically significant indicating that the adoption of plant management was higher in those districts where the SRI was not incorporated under the NFSM scheme.

Additionally, unlike many previous studies on agricultural technology adoption, the education of the household was not significant in affecting the adoption decisions. This is due to the fact that unlike other innovations, SRI is a farmer centric innovation and the success of it depends largely on farmers' motivation and skills rather than the level of education. But surprisingly, the access to extension services reported by farmers did not have any impact on the adoption decisions. Perhaps this is somewhat linked with the lack of impact of NFSM on SRI adoption. The ineffectiveness of extension services would be the reason for the lack of impact of NFSM on the adoption of SRI.

On the impact of adoption of SRI on welfare outcomes, the results showed the sample selection bias especially if the income equation is estimated without considering the adoption decision. The results between exogenous and endogenous adoption decisions showed differences mainly in estimates for household income, whereas in the case of yield almost all combinations of SRI had a positive and statistically significant impact on yield under the exogenous and endogenous assumptions. However, in the case of household income, the positive and statistically significant impact of income was evident in the adoption of two combinations/practices—plant management plus soil management and plant management plus water management. However, the adoption of plant management practice and the adoption of plant plus water plus soil management had a negative and statistically significant impact on income.

The impact estimates also highlight the fact that the NFSM had a positive and statistically significant impact on rice yield and income, and this indicates that the rice yield and household income were relatively higher in those districts where the SRI promotion had been incorporated under the NFSM. But it is clear that the positive impact of NFSM on rice yield and household income was not due to the increased adoption of SRI. As noted earlier, NFSM did not have a positive impact in promoting the adoption.

As far as the policy implications are concerned, the results suggest that even the partial adoption of SRI principles can have an impact on rice yield. The impact of SRI adoption on income was mixed. However, the results indicate that the SRI adoption can improve the household income provided the government intervenes in reducing the cost involved in the adoption of some of practices/components of SRI. For example, supply of mechanical weeders can ensure more efficient weeding and thereby full realisation of the benefits. Similarly, skill-oriented training to agricultural labourers can also help in reducing the transaction costs involved in the adoption of SRI practices. There is an urgent need for extension services to be scaled up by making these services more effective. Small farmers were found to be more enthusiastic in adopting SRI than large farmers. Given the limited amount of resources that the small and marginal farmers possess, there is an urgent need for government to intervene to make extension services more effective and thereby to make a positive impact of NFSM on the adoption of SRI. Briefly, adoption of different practices/combinations of SRI can boost yield and household income with appropriate policy interventions.

Appendix

See Tables 9.4 and 9.5.

Table 9.4 Variable definitions

Variable	Definition
No. of family members	Number of the family members in the household including children
Active family labourers	Number of active family labourers
Education	Number of family members who have education higher than tenth class
Size of the landholding	Total size of owned and rented landholdings cultivated by household in hectares
Assets owned	Total number of the type of farm assets (bullocks, tractors, machinery, etc.) owned by the household
Membership in farmer organisations	Membership of any of the family member in farmer organisations, dummy variable = 1 if any of the family member has membership in farmer organisations, = 0 otherwise
Irrigation facility	Dummy variable = 1 if the household had irrigation sources other than rainfall, = 0 otherwise
Access to off-farm activity	Dummy variable = 1 if the household had access to off-farm activity, = 0 otherwise
Improved varieties known	Number of improved rice varieties known to the household
Terrain type	Dummy variable = 1 if the terrain is levelled, = 0 otherwise
Distance to main market	Distance to the nearest main market (in kilometres)
Fear of poor yield	Fear of poor rice yield due to the adoption of SRI/any new method of rice cultivation. Dummy variable = 1 if the household has fear of poor yield due to SRI adoption, = 0 otherwise
Experience in agriculture	Number of years the household has been cultivating rice
Access to extension services	Dummy variable = 1 if the household had access to extension services from NGOs or government departments regarding SRI practices including training and guidance, = 0 otherwise
Have information about SRI	Dummy variable = 1 if the household had information regarding SRI from agricultural departments and NGOs, = 0 otherwise
NFSM dummy	Dummy variable = 1 for those districts where SRI is incorporated within NFSM, = 0 otherwise

Table 9.5 Parameter estimates: test on validity of selection instruments

Variables	Ln rice yields/ha	Ln household income
No. of family members	−0.01 (0.02)	−0.03 (0.01)***
Active family labourers (in no.)	0.03 (0.02)	0.03 (0.01)**
Education	0.06 (0.02)***	0.06 (0.01)***
Size of the landholding	−0.02 (0.01)*	0.05 (0.01)***
Assets owned	−0.03 (0.02)	0.02 (0.03)
Irrigation facility	0.05 (0.05)	0.03 (0.03)
Access to off-farm activity (Yes = 1)	0.14 (0.05)***	0.07 (0.03)**

(continued)

Table 9.5 (continued)

Variables	Ln rice yields/ha	Ln household income
Membership in input supply cooperative	−0.05 (0.06)	0.06 (0.05)
Improved varieties known	0.05 (0.03)*	0.10 (0.02)***
Terrain type	0.19 (0.07)***	0.17 (0.04)***
Fear of poor yield	−0.06 (0.06)	−0.08 (0.03)**
Experience in agriculture	0.00 (0.00)***	−0.00 (0.00)
Contact extension service	0.13 (0.07)*	−0.02 (0.05)
Distance to main market	0.09 (0.03)***	0.01 (0.02)
NFSM dummy	0.13 (0.04)***	0.04 (0.03)
Information from agricultural department	0.07 (0.06)	−0.05 (0.04)
Constant	0.42 (0.14)***	−0.58 (0.09)***
N	193	193

Note ***$p < 0.01$, **$p < 0.05$, *$p < 0.1$. Robust standard errors are given in parentheses

Chapter 10
Conclusion and Policy Implications

Rice plays a major role in meeting the food requirements of half of the world's population. Around 90% of the rice produced is consumed in the Asian region, and for people who live in Asia, rice security is equivalent to food security. As a major producer and consumer, India plays an important role in the global rice economy. Nonetheless, rice cultivation in India today suffers from several interrelated problems. Increased productivity achieved during the green revolution period through input-intensive methods of high water and fertiliser use in well-endowed regions is showing signs of stagnation and concomitant environmental problems due to salinisation and waterlogging of fields. Since virtually all suitable land is already under cultivation, raising productivity—through simultaneous reduction of negative environmental consequences and improvement of the efficient use of resources—seems to be the only way forward.

SRI, which originated in Madagascar, is widely recognised as a promising systemic approach to rice production by reducing negative environmental effects. SRI was widely believed to enhance yield and substantially reduce water and other input requirements (Stoop et al. 2002; Uphoff 2002, 2003) through changes in the management of plants, soil, water and nutrients (Satyanarayana et al. 2007).

SRI is not a technology but a different method of cultivation. Unlike other technological innovations, SRI does not require costly investment. So, SRI is a knowledge-based innovation. Therefore, one would naturally expect SRI to be widely disseminated and successfully adopted. Despite the potential benefits of SRI, its adoption rate is very low and also varies from region to region. Studies point out factors such as poor water control, lack of awareness, skill-intensive nature of the method, difficulty in getting labourers, etc. as constraints in adoption. These could be the reasons for the common practice of partial adoption of components observed in most of the regions that adopted SRI. These constraints are even more severe in a developing country like India. Against this backdrop, the present study analysed the factors influencing the adoption of SRI as well as the impact of SRI adoption on household income and rice yield. The study made the analysis from three different perspectives. First was the analysis of the factors

P. Varma, *Rice Productivity and Food Security in India*,
DOI 10.1007/978-981-10-3692-7_10

influencing the intensity and depth of adoption by explicitly considering effective information constraints in the adoption of SRI. Second was the analysis of the factors influencing the adoption of various components of SRI and the combinations of various components of SRI. Third was the analysis of the adoption and impact of SRI on income and rice yield in a joint framework.

This study has been divided into 10 chapters including introduction and conclusion. Chapter 1 as an introduction provided the background, objectives, data, methodology and the scope of the study along with chapter scheme. Chapter 2 gave an overview of rice economy. Chapter 3 discussed the role of sustainable rice production practices to improve the rice productivity and food security with an emphasis on India. Chapter 4 discussed the SRI in detail by discussing the experiences at international, national and farm levels. Chapter 5 discussed the salient features of government of India's National Food Security Mission (NFSM) and its objectives especially in the context of rice production and SRI. Chapter 6 provided an overview of SRI adoption across different study regions along with a detailed discussion of the socio-economic profile of the sample households. Chapter 7 analysed the factors that influence the adoption of SRI under effective information constraints. Chapter 8 provided an in-depth understanding of factors that influence the adoption of multiple practices of SRI. Chapter 9 analysed the impact of SRI adoption on household income and rice yield. This chapter provided the conclusion and policy implications of the study.

The detailed household data analysis was conducted for 3 major rice-producing states. They are Karnataka, Orissa and Madhya Pradesh. For the purpose of further identification of the districts, the study made use of three criteria, viz. agro-climatic zones, SRI districts and SRI-NFSM districts. Thus, 2 districts from each state that belong to the same agro-climatic zones were identified. Selection of the districts in each state belonging to same agro-climatic zones was done after ensuring that one is with SRI practice and the other is with SRI incorporated under NFSM.

For Karnataka, Hassan (SRI-NFSM) and Chikmagalur (SRI) districts are identified. Similarly for Orissa, Keonjhar (SRI-NFSM) and Mayurbhanj (SRI) districts are identified. For Madhya Pradesh, Sidhi (SRI) and Shahdol (SRI-NFSM) districts are selected. After selection of districts, the blocks/taluks were selected from each district.

Primary data was collected through a comprehensive household survey in the above-mentioned six districts of three major rice-producing Indian states in the year 2015. The farmers were selected through a stratified random sampling technique. The sample ensured the representation of almost all the available SRI farmers from the study region and equal number of non-adopters of SRI. The survey was conducted through questionnaire, framed in such way as to draw out details covering household characteristics, wealth and farm characteristics, institutional and access-related variables, risk and economic factors. The precise questions or variables included in the questionnaire were based on the insights received from various studies on agricultural technology adoption. The total number of farm households interviewed was 386, of which 193 households were SRI adopters.

After discussing the background, objectives, data and methodology in the first chapter, the second chapter provided an overview of Asian rice economy with a special emphasis on Indian rice economy. The chapter discussed the recent performance of India by discussing the trends in rice production and productivity of rice in the states as well as the districts selected for the purpose of analysis.

The third chapter discussed the major rice production technologies by discussing the importance of sustainable agricultural practices in general and in the context of rice cultivation in particular. The fourth chapter provided a detailed discussion of SRI and the role of institutions in promotion of SRI, etc. Some of the scepticisms associated with SRI were also discussed. The farmer's experience of adoption of SRI based on our field-level observations pointed out that most farmers agree with the higher yield benefits of SRI as compared to traditional rice cultivation. The farmers also opined that they require less amounts of inputs for SRI cultivation. However, the major lacuna in adoption pointed out by farmers in the study areas was the lack of availability of labourers for both SRI and rice cultivation. This was a major cause for concern in all the study areas, which was even more pronounced in Karnataka. The reason for less difficulty in getting labourers for rice cultivation in some other study regions was due to the availability of family labourers for farming. For example, in districts such as Shahdol, the labour issues were minimal due to the fact that the farming was heavily dependent upon the family labourers. Another interesting observation from farmers' experience was the difficulty in adopting water management—alternate wetting and drying—due to heavy rainfalls. So those farmers who were dependant on rainfall for rice cultivation were not able to adopt the water management practices even when they adopted other practices of SRI. Therefore, partial adoption was common. In some districts like Shahdol, inadequate rainfall or delay in rainfall was also a major cause of concern. Another major lacuna in adoption was the lack of equipment such as cono-weeder which discouraged the farmers from adopting cono-weeders for weeding which is also an important component of SRI. The government's support and subsidy for equipment such as cono-weeder may not be sufficient; so, the initial observations pointed out the urgent need to scale up these subsidies.

The preliminary analysis of the data was undertaken in the sixth chapter showed that majority of farm households were poor or marginal with farming or wage labour as their main sources of income. Additionally, majority of the households interviewed were headed by males. Interestingly, lack of information about SRI was not a major issue as only 10% of farmers lacked information regarding SRI. On the other hand, 53% of total farmers and 63% of non-adopters did not have access to extension services clearly indicating that lack of training and guidance in terms of adoption were bigger concerns. Non-availability of labour was reported by most farmers—a major concern even among adopters of SRI. Similarly, 40% of adopters and 60% of non-adopters solely relied upon rainfall as the major source of irrigation indicating that lack of irrigation was a likely reason for non-adoption. There were huge variations in socio-economic profiles of farmers across different districts as well. The analysis also showed that in NFSM districts, the main source of information about SRI was the government's agricultural departments, whereas in non-NFSM districts the main source of information about SRI was the NGOs. This shows that SRI is taking roots in India through diverse channels. The analysis based on the descriptive statistics of yield, household income and cost of cultivation showed that the adopters of SRI had higher yield, income and cost benefits. Although the analysis showed that adoption of some components like soil management is more costly than others, the adoption of SRI in general appears to be beneficial for rice farmers. The remaining chapters undertook rigorous empirical analysis to substantiate our observations and field insights.

The seventh chapter analysed the determinants of adoption of SRI by using two kinds of simultaneous equation models. The first set of simultaneous equation models analysed the status of SRI adoption (yes/no) in the first stage and intensity (depth) of SRI adoption in the second stage, whereas the second set of simultaneous equation models explicitly incorporated the effective information constraints in the model. When farmers lack effective information about a new practice, even farmers with a positive desired demand will not be able to adopt the new practice. Therefore, the seventh chapter developed a simultaneous equation model by explicitly incorporating access to effective information as a constraint in adoption.

The results from the analysis offer unique policy-relevant insights on the importance of effective information in conditioning the intensity as well as the depth of SRI adoption in India. Even when around 90% of the sample farmers had information about SRI, only 76% of sample farmers received information about SRI from formal sources such as Agricultural departments and NGOs. The argument is further strengthened by the fact that among the adopters around 96% of farmers received information from Agricultural departments and NGOs whereas among the non-adopters only 55% received information from these sources. The effective information is therefore deemed to be important in influencing the final adoption decision of farmers.

The information access model showed that none of the household- or farm-level variables were statistically significant. Although variables like education, size of the landholding, household size, farm assets owned, etc. showed positive sign, none of these variables were statistically significant. The farmers' social capital variables like membership in farmer organisations and seed exchange experience were positive and statistically significant. The results indicate the importance of social capital of the farmers in accessing greater information. The information and communication variable (ICT), ownership of cell phone, had a positive and statistically significant impact on accessing effective information.

The dummy variable, NFSM, to capture the differences in information available in NFSM districts and non-NFSM districts was positive but statistically insignificant. Therefore, the NFSM did not have a statistically significant impact in ensuring higher effective information about SRI to farmers.

The results for intensity of adoption of SRI as well as the depth of SRI adoption further strengthened the importance of social capital variable—membership in farmer organisation—in the adoption of SRI. The results from both the models showed that younger farmers were more enthusiastic in adopting SRI than older farmers. As hypothesised, availability of labour and effective information was significant in influencing the intensity of adoption. The irrigation facility did not come out to be significant in the adoption of SRI in terms of acres indicating the importance of irrigation and irrigation management only in certain components of SRI, i.e., alternate wetting and drying.

More number of variables were significant in affecting the depth of adoption model than intensity of adoption model. This is not surprising as many factors (e.g. irrigation facility as mentioned earlier and the cost of labourers) will appear to be more relevant when farmers adopt greater number of SRI practices. Irrigation facility was significant and positive in affecting the depth of adoption. Assets owned were also significant and positive. Farm assets such as mechanical hand weeders are more relevant only when farmers decide to adopt the SRI component of mechanical hand weeders to aerate the soil. Similar to intensity model, availability of labourers and access to effective information came out to be positive and significant in the depth of adoption model as well.

Wage rates for labourers were crucial in determining the depth of adoption of SRI. Wage rates of woman labourers were negatively related to adoption, whereas wage rates for male labourers were positively related to adoption. This is perhaps due to the fact that the shift away from manual weeding to mechanical weeding creates more demand for male labourers. So, the skill-intensive nature of mechanical weeding is leading to higher demand for male labourers and thus higher wages. It points to the possibility of a gender-biased technical change. The results therefore indicated the skill-intensive nature of SRI adoption and the gender implications of SRI adoption. Although the NFSM dummy variable had a positive sign in all the models, the variable was significant neither in the information models nor in the adoption models.

In Chap. 8, we analysed the probability and level of adoption of multiple practices of SRI by rice farmers of selected districts. We used MVP and ordered probit models to jointly analyse the adoption of multiple practices and the number of SRI practices adopted while recognising the interrelationship among them. Our approach extends the existing empirical studies by allowing for correlations across different practices of SRI. The results showed that various economic, institutional and access-related factors shape farmers' adoption of SRI practices.

The analysis found a significant and positive relationship between households' assets and adoption of SRI practices. Similar to the simultaneous equation models in Chap. 7, the wage rates of women labourers had a negative and significant impact, whereas the wage rates of male labourers had a significant and positive impact on adoption of SRI. Therefore, the results further strengthened the fixed social bias and gender disparities in the adoption of SRI. The results are very much in line with the results that we obtained for the earlier simultaneous equation model on the depth of adoption of SRI.

Information and extension services are very important driving forces for enhancing adoption of SRI. Knowledge and awareness are particularly relevant for innovations such as SRI where the adoption is more from sustainability perspective than immediate benefits (Noltze et al. 2012). Our results showed the importance of extension services in influencing adoption decisions. Similar to the simultaneous equation model analysis in Chap. 7, the dummy variable for NFSM was insignificant in the analysis of the adoption of multiple practices as well. The insignificant impact of NFSM districts dummy on SRI adoption is an eloquent testimony to the fact that the objective of increased rice production by promoting SRI under the government's Food Security Mission did not yield the desired results. Additionally, farmers who were remotely located from the main market had higher likelihood of adopting SRI. This also indicates the possibility for cultivation of commercial crops by those farmers who have better proximity to the market. Most farmers who we interviewed do not consider rice farming as a commercially viable venture; instead, the production was mainly for self-consumption and sale in the local markets.

Similar to the analysis based on simultaneous equation models, the analysis in Chap. 8 also pointed out the importance of investment in infrastructure such as irrigation in promoting SRI. The significance of social capital and networks was also observed in our analysis. As in the case of analysis conducted in Chap. 8, the membership in farmers' organisations such as input supply cooperatives increased the likelihood of different SRI adoption practices.

Chapter 9 examined the determinants and impacts of the adoption of five mutually exclusive combinations of SRI on yields and household incomes using a multinomial endogenous treatment effects model. As in the models explained earlier, we find that the decision to adopt is a function of farm assets, irrigation facility, information about SRI from formal sources (effective information as defined in Chap. 7), experience in agriculture, number of improved rice varieties known, access to off-farm activity, etc. As in the case of other models, the number of farm assets came out to be significant in the adoption of all SRI practices/combinations. The variable for information was positive and statistically significant in the adoption of three out of four SRI practices/combinations. Therefore, the results further strengthened the importance of effective information in the adoption. As in the case of the analysis undertaken in Chaps. 7 and 8, irrigation facility came out to be significant in the adoption decisions even when we analysed the adoption and impact in a joint framework. The variable for irrigation facility came out to be significant in the adoption of three out of four SRI practices/combinations. In the case of plant and soil management practices, the irrigation facility had a negative and statistically significant impact. This indicates that the higher adoption of plant and soil management practices by households was due to the lack of proper irrigation facility.

Apart from the above factors, the other variables that affected the adoption of some of the practices/combinations of SRI were number of improved rice varieties known, household size, access to off-farm activity, fear of poor yield, distance to main market, type of terrain, size of the landholding and experience in agriculture. Number of improved rice varieties known, household size, type of terrain (terrain which is levelled), etc. increased the likelihood of household adopting different practices/combinations of SRI, whereas experience in agriculture or the number of years the farmer stayed in rice cultivation, access to off-farm activity, size of the landholding, etc. decreased the likelihood of adopting different practices/combinations of SRI. The dummy variable for NFSM was insignificant in most cases. In the case of the adoption of plant management, the NFSM was negative and statistically significant indicating that the adoption of plant management was higher in those districts where the SRI was not incorporated under the NFSM scheme.

On the impact of adoption of SRI on welfare outcomes, the results showed the sample selection bias especially if the income equation is estimated without considering the adoption decision. The results between exogenous and endogenous adoption decisions showed differences mainly in estimates for household income, whereas in the case of yield almost all combinations of SRI had a positive and statistically significant impact on yield under the exogenous and endogenous assumptions. However, in the case of household income, the positive and statistically significant impact of income was evident in the adoption of two combinations/practices—plant management plus soil management and plant management plus water management. However, the adoption of plant management practice and the adoption of plant plus water plus soil management had a negative and statistically significant impact on income.

The impact estimates also highlight the fact that the NFSM had a positive and statistically significant impact on rice yield and income, and this indicates that the rice yield and household income were relatively higher in those districts where the SRI promotion had been incorporated under the NFSM. But it is clear that the positive impact of NFSM on rice yield and household income was not due to the increased adoption of SRI. As noted earlier, NFSM did not have a positive impact in promoting the adoption.

Briefly put, the results from various econometric analyses complemented each other. The analysis based on various econometric exercises was quite useful in better understanding the welfare benefits of SRI adoption and the plethora of constraints that the farmers face in adoption.

As far as the policy implications are concerned, the results suggest that even the partial adoption of SRI principles can have an impact on rice yield. The impact of SRI adoption on income was mixed. However, the results indicate that the SRI adoption can improve the household income provided the government intervenes in reducing the cost involved in the adoption of some of the practices/components of SRI. For example, supply of mechanical weeders can ensure more efficient weeding and thereby full realisation of the benefits. Similarly, skill-oriented training to agricultural labourers can also help in reducing the transaction costs involved in the adoption of SRI practices. There is an urgent need for extension services to be scaled up by making these services more effective. Small farmers were found to be more enthusiastic in adopting SRI than large farmers. Given the limited amount of resources that the small and marginal farmers possess, there is an urgent need for government to intervene to make extension services more effective and thereby to make a positive impact of NFSM on the adoption of SRI.

The results pointed out the lacuna in government intervention in disseminating the awareness and the merits of SRI among farmers. The NFSM districts were not significantly different in disseminating the information about SRI as understood from the analysis. This could be the reason why NFSM did not have a positive impact on adoption of SRI. Therefore, the study points out the urgent need for scaling up of the programmes to encourage more and more farmers to adopt SRI. The study also shows the importance of irrigation facility, so there is a need for major investments in upgrading the irrigation scheme. Although SRI requires less water as compared to traditional method, farmers require their own irrigation resources to adopt proper water management which is an essential component of SRI.

The findings also revealed that the adoption of SRI is at the expense of women employment. Therefore, gender implications of SRI adoption need to be evaluated as a matter of some priority. The capacity building and targeted training is highly required not just for farmer groups but also for women agricultural labourers to upgrade their skills and expertise in using some of the farm machineries especially the mechanical weeders.

The significance of membership in farmer organisations, input supply cooperatives, etc. indicates that that policymakers should focus on establishing and strengthening local collective institutions. Local institutions can play a crucial role in providing farmers with timely information, inputs and technical assistance. Unlike many previous studies on agricultural technology adoption, the education of the household was not significant in affecting the adoption decisions. This is due to the fact that unlike other innovations, SRI is a farmer-centric innovation, and the success of it depends largely on farmers' motivation and skills rather than the level of education. Therefore, greater government intervention is needed to enhance farmer's motivation and the understanding of the environmental and economic benefits of the SRI adoption.

To sum it up, the wider adoption of SRI can contribute to promoting not only sustainable agricultural practices but also greater food security provided the constraints that the farmers are facing are addressed with appropriate policy interventions.

References

Abdulai, A., and W. Huffman. 2014. The adoption and impact of soil and water conservation technology: An endogenous switching regression application. *Land Economics* 90 (1): 26–43.

Abraham, Binju, Hailu Araya, Tareke Berhe, Sue Edwards, Biksham Gujja, Ram Bahadur Khadka, Yang S Koma, Debashish Sen, Asif Sharif, Erika Styger, Norman UphoffEmail author and Anil Verma. 2014. The system of crop intensification: reports from the field on improving agricultural production, food security, and resilience to climate change for multiple crops. *Agriculture & Food Security*. doi:10.1186/2048-7010-3-4.

Adegbola, P., and C. Gardebroek. 2007. The effect of information sources on technology adoption and modification decisions. *Agricultural Economics* 37 (1): 55–65.

Adesina, A.A., and M.M. Zinnah. 1993. Technology characteristics, farmers' perceptions and adoption decisions: A Tobit model application in Sierra Leone. *Agricultural economics* 9 (4): 297–311.

Agro-economic research centre for Madhya Pradesh and Chhattisgarh. 2013. Spread of new varieties of hybrid rice and their impact on the overall production and productivity in Madhya Pradesh. Available at http://www.jnkvv.nic.in/images/AERC%202013/Report/108%20Hybrid%20Rice.pdf.

Asfaw, S., B. Shiferaw, F. Simtowe, and M.G. Haile. 2010. Agricultural technology adoption, seed access constraints and commercialization in Ethiopia. *Journal of Development and Agricultural Economics* 3 (9): 436–477.

Aune, J. B., N.U. Sekhar, K. Esser, and M. Tesfai. 2014. Opportunities for support to system of rice intensification in Tanzania, Zambia and Malawi. Noragric Report No. 71. Available at https://www.nmbu.no/sites/default/files/pdfattachments/noragric_report_no._71.pdf.

Awotide, B.A., A. Diagne, A.N. Wiredu, and V.E. Ojehomon. 2012. Wealth status and agricultural technology adoption among smallholder rice farmers in Nigeria. *OIDA International Journal of Sustainable Development* 5 (2): 97–108.

Bansal, S., and P.S. Ward. 2015. *Do farmers value rice varieties tolerant to droughts and floods?*. India: Evidence from a discrete choice experiment in Odisha.

Barah, B. 2009. Economic and Ecological benefits of system of rice intensification (SRI) in Tamil Nadu. *Agricultural Economics Research Review* 22 (2): 209–214.

Barison, J., and A. Kassam. 2011. The system of rice intensification as a beneficial human intervention into root and soil interaction. *Jurnal Tanah Lingkungan* 13 (2): 72–88.

Barrett, C. B., et al. 2016. System of rice intensification in rural Bangladesh: Adoption, diffusion and impact. Working Paper, International Growth Centre.

Barrett, C.B., C.M. Moser, O.V. McHugh, and J. Barison. 2004. Better technology, better plots or better farmers? Identifying changes in productivity and risk among Malagasy rice farmers. *American Journal of Agricultural Economics* 86 (4): 869–889.

Bassey, J.I. 2016. Comparative analysis of system of rice intensification and traditional system of rice production in Abi LGA, Cross River State, Nigeria. *European Journal of Agriculture and Forestry Research* 4 (2): 9–23.

P. Varma, *Rice Productivity and Food Security in India*,
DOI 10.1007/978-981-10-3692-7

Basu, S., and C. Leeuwis. 2012. Understanding the rapid spread of system of rice intensification (SRI) in Andhra Pradesh: Exploring the building of support networks and media representation. *Agricultural Systems* 111: 34–44.

Bhatt, K.N. 2015. *System of rice intensification for increased productivity and ecological security: A report* 3: 147. doi: 10.4172/2375-4338.1000147.

Bishwajit, G., S. Sarker, M.A. Kpoghomou, H. Gao, L. Jun, D. Yin, and S. Ghosh. 2013. Self-sufficiency in rice and food security: A South Asian perspective. *Agriculture & Food Security* 2 (1): 1.

Central Rice Research Institute. 2011. *Vision 2030*. Cuttack, Orissa: Indian Council of Agricultural Research.

Chadhar, A.R., M.A. Nadeem, A. Tanveer, and M. Yaseen. 2014. Weed management boosts yield in fine rice under system of rice intensification. *Planta Daninha* 32 (2): 291–299.

Chowdhury, M.R., V. Kumar, and K. Brahmachari. 2012. Effect of different water and nutrient management practices on rice grown under SRI. *Journal of Crop and Weed* 8 (2): 36–39.

Ches, S., and E. Yamaji. 2016. Labor requirements of system of rice intensification (SRI) in Cambodia. *Paddy and Water Environment* 14 (2): 335–342.

Croppenstedt, A., M. Demeke, and M.M. Meschi. 2003. Technology adoption in the presence of constraints: the case of fertilizer demand in Ethiopia. *Review of Development Economics* 7: 58–70.

Dar, M.H., S. Singh, U.S. Singh, N.W. Zaidi, and M.M. Ismail. 2014. Stress tolerant rice varieties-making headway in India. *SATSA Mukhaptra Annual Technical Issue* 18: 1–14.

Das, S.R. 2012. Rice in Odisha. *IRRI Technical Bulletin* (16).

Davis, B., P. Winters, T. Reardon, and K. Stamoulis. 2009. Rural nonfarm employment and farming: Household-level linkages. *Agricultural Economics* 40 (2): 119–123.

Deb, P., and P.K. Trivedi. 2006. Maximum simulated likelihood estimation of a negative binomial regression model with multinomial endogenous treatment. *Stata Journal* 6 (2): 246–255.

Deb, P., and P.K. Trivedi. 2006. Specification and simulated likelihood estimation of a non-normal treatment-outcome model with selection: Application to health care utilization. *The Econometrics Journal* 9 (2): 307–331.

Deb, P. 2009. MTREATREG: Stata module to fits models with multinomial treatments and continuous, count and binary outcomes using maximum simulated likelihood. *Statistical Software Components*.

Diagne, A., and M. Demont. 2007. Taking a new look at empirical models of adoption: Average treatment effect estimation of adoption rates and their determinants. *Agricultural Economics* 37 (2–3): 201–210.

Di Falco, S., and E. Bulte. 2011. A dark side of social capital? Kinship, consumption, and savings. *Journal of Development Studies* 47 (8): 1128–1151.

Di Falco, S., and M. Veronesi. 2013. How can African agriculture adapt to climate change? A counterfactual analysis from Ethiopia. *Land Economics* 89 (4): 743–766.

Dill, J., G. Deichert and Le Thi Nguyet Thu. 2013. Promoting the system of rice intensification-lessons learned from Tra Vinh Province, Vietnam, IFAD. Available at: http://infoagro.net/archivos_Infoagro/Regatta/biblioteca/VN-GIZreportonLesson.pdf.

Directorate of Agriculture & Food Production. (2011–12). *Odisha agriculture statistics*. Government of Odisha.

Dobermann, A. 2004. A critical assessment of the system of rice intensification (SRI). *Agricultural Systems* 79 (3): 261–281.

Dorfman, J.H. 1996. Modeling multiple adoption decisions in a joint framework. *American Journal of Agricultural Economics* 78 (3): 547–557.

Doss, C.R., and M.L. Morris. 2001. How does gender affect the adoption of agricultural innovations? The case of improved maize technology in Ghana. *Agricultural Economics* 25 (1): 27–39.

FAOSTAT. 2014. Food and agriculture organization of the United Nations Statistics Division. Available at: http://faostat3.fao.org/.

Feder, G. 1982. Adoption of interrelated agricultural innovations: Complementarity and the impacts of risk, scale, and credit. *American Journal of Agricultural Economics* 64 (1): 94–101.

Feder, G., R.E. Just and D. Zilberman. 1985. Adoption of agricultural innovations in developing countries: A survey. *Economic development and cultural change*, 255–298.

Gathorne-Hardy, A., D.N. Reddy, M. Venkatanarayana, and B. Hariss-White. 2016. System of rice intensification provides environmental and economic gains but at the expense of social sustainability: A multi-disciplinary analysis in India. *Agricultural Systems* 2016 (143): 159–168.

Geethalakshmi, V., A. Lakshmanan, D. Rajalakshmi, R. Jagannathan, G. Sridhar, A.P. Ramaraj, K. Bhuvaneswari, L. Gurusamy, and R. Anbhazhagan. 2011. Climate change impact assessment and adaptation strategies to sustain rice production in Cauvery basin of Tamil Nadu. *Current Science* 101 (3): 342–347.

Glover, D., et al. 2016. The adoption problem; or why we still understand so little about technological change in African agriculture. *Outlook on Agriculture* 45 (1): 3–6.

Glover, D. 2011. Science, practice and the system of rice intensification in Indian agriculture. *Food Policy* 36 (2011): 749–755.

Government of Orissa. 2014–15. Activity Report of Agriculture Department. 2014–15.

Gujja, B., and T.M. Thiyagarajan. 2009. *New Hope for Indian food security? The system of rice intensification*. International Institute for Environment and Development.

Gujja, B., and T.M. Thiyagarajan. 2013. *New Hope for Indian food security? The system of rice intensification*. The Gatekeeper Series of the Natural Resources Group at IIED. Available at: http://pubs.iied.org/pdfs/14587IIED.pdf. Accessed on August 22, 2014.

Gujja, B., N. Loganandhan, and V.V. Goud. 2008. *System of rice intensification: Experiences of farmers in India*. Andhra Pradesh, India: ICRISAT-WWF Project.

Gujja, B., and N. Uphoff. 2013. From animosity towards collaboration. *Agricultures Network*. Available at: http://www.agriculturesnetwork.org/news/gujja-uphoff-respond-bouman.

Haldar, S., Honnaiah and G. Govindaraj. 2012. *System of rice intensification (SRI) method of rice cultivation in West Bengal (India)*. An Economic analysis, Selected Poster prepared for presentation at the International Association of Agricultural Economists (IAAE) Triennial Conference, Foz do Iguaçu, Brazil, August 18–24, 2012. Available at: http://ageconsearch.umn.edu/bitstream/126234/2/IAAE%20conference%20paper%20on%20SRI-uploading.pdf. Accessed on November 16, 2015.

Howell, K.R., P. Shrestha, and I.C. Dodd. 2015. Alternate wetting and drying irrigation maintained rice yields despite half the irrigation volume, but is currently unlikely to be adopted by smallholder lowland rice farmers in Nepal.*Food and Energy Security* 4 (2): 144–157.

IIED, Available at: http://pubs.iied.org/pdfs/14587IIED.pdf. Accessed on November 2, 2015.

International Food Policy Research Institute, IFPRI. 2009. *Millions fed: Proven successes in agricultural development*. Washington, DC: International Food Policy Research Institute (IFPRI).

IRRI. 2012–13. *Cluster demonstrations of stress tolerant rice varieties in stress prone parts of India*. Annual Report submitted to National Food Security Mission, Ministry of Agriculture, Government of India.

Jha, D.K. 2014. New FAO norms comes as a shot in the arms of rice exporters. *Business Standard*. http://www.business-standard.com/article/markets/new-fao-norms-come-as-a-shot-in-the-arm-for-rice-exporters-114072601087_1.html.

Johnson, B., and K. Vijayaragavan. 2011. Diffusion of System of Rice Intensification (SRI) Across Tamil Nadu and Andhra Pradesh in India. *The Indian Research Journal of Extension Education* 11 (3): 72–79.

Jones, A. 1989. A double-hurdle model of cigarette consumption. *Journal of Applied Econometrics* 4 (1): 23–39.

Joshi, P.K., D. Roy, V. Sonkar and G. Tripathi. 2016. Technologies for maize, wheat, rice and pulses in marginal districts of Bihar and Odisha. In *Technological and institutional innovations for marginalized smallholders in agricultural development*, eds. F.W. Gatzweiler and J.V. Brown. Springer.

Just, R.E., and D. Zilberman. 1983. Stochastic structure, farm size and technology adoption in developing agriculture. *Oxford Economic Papers*, 307–328.

Kabunga, N.S., T. Dubois, and M. Qaim. 2012a. Heterogeneous information exposure and technology adoption: The case of tissue culture Bananas in Kenya. *Agricultural Economics* 43: 473–486. DOI: 10.1111/j.1574-0862.2012.00597.x.

Kabunga, Nassul S., Thomas Dubois, and Matin Qaim. 2012b. Yield effects of tissue culture bananas in Kenya: Accounting for selection bias and the role of complementary inputs. *Journal of Agricultural Economics* 63 (2): 444–464. DOI: 10.1111/j.1477-9552.2012.00337.x.

Kassie, M., H. Teklewold, P. Marenya, M. Jaleta and O. Erenstein . 2014. Production risks and food security under alternative technology choices in Malawi: Application of a multinomial endogenous switching regression. *Journal of Agricultural Economics.*

Kassie, M., M. Jaleta, B. Shiferaw, F. Mmbando, and M. Mekuria. 2013. Adoption of interrelated sustainable agricultural practices in smallholder systems: Evidence from rural Tanzania. *Technological Forecasting and Social Change* 80 (3): 525–540.

Kassie, M., H. Teklewold, M. Jaleta, P. Marenya, and O. Erenstein. 2015. Understanding the adoption of a portfolio of sustainable intensification practices in eastern and southern Africa. *Land Use Policy* 42: 400–411.

Kathage, J., M. Kassie, B. Shiferaw, and M. Qaim. 2016. Big constraints or small returns? Explaining nonadoption of hybrid maize in Tanzania. *Applied Economic Perspectives and Policy* 38 (1): 113–131. DOI: 10.1093/aepp/ppv009.

Khanal, N. P., and K.L. Maharjan. 2014. Adoption of improved rice varieties in the Tarai region of Nepal. In *Communities and Livelihood Strategies in Developing Countries*, 49–64. : Springer.

Khandker, V. and V.P. Gandhi. 2012. Introduction of new technologies in agriculture: A study of the challenges in the adoption of hybrid rice in India. In *Paper prepared for presentation at the 56th AARES annual conference*, Fremantle, Western Australia, February 7–10, 2012. Available at: http://ageconsearch.umn.edu/bitstream/124325/2/2012AC%20Khandker%20CP.pdf. Accessed on October 1, 2015.

Khonje, M., J. Manda, A.D. Alene, and M. Kassie. 2015. Analysis of adoption and impacts of improved maize varieties in eastern Zambia. *World Development* 66: 695–706.

Kumar, A., and A. Saha. 2013. Engagement of the state in the promotion of SRI: Understanding the process. *News Reach Magazine* 13(2): 31–49.

Kumar, V., and J.K. Ladha. 2011. 6 Direct seeding of rice: Recent developments and future research needs. *Advances in Agronomy* 111: 297.

Lakkakula, P., B.L. Dixon, M.R. Thomsen, E.J. Wailes, and D.M. Danforth. 2015. Global rice trade competitiveness: a shift-share analysis. *Agricultural Economics* 46 (5): 667–676.

Lampayan, R.M., R.M. Rejesus, G.R. Singleton, and B.A. Bouman. 2015. Adoption and economics of alternate wetting and drying water management for irrigated lowland rice. *Field Crops Research* 170: 95–108.

Langyintuo, A.S., and C. Mungoma. 2008. The effect of household wealth on the adoption of improved maize varieties in Zambia. *Food policy* 33 (6): 550–559.

Leesawatwong, M., S. Jamjod, J. Kuo, B. Dell, and B. Rerkasem. 2005. Nitrogen Fertilizer increases seed protein and milling quality of rice. *Cereal Chemistry* 82: 588–593.

Lidia, Dandedjrohoun, Aliou Diagne, Gauthier Biaou, Simon N'Cho, and Soul-Kifouly Midingoyi. 2012. Determinants of diffusion and adoption of improved technology for rice parboiling in Benin. *Revue d'Etudes en Agriculture et Environnement - Review of agricultural and environmental studies, INRA Editions* 93 (2): 171–191.

Ly, P., L.S. Jensen, T.B. Bruun, D. Rutz, and A. de Neergaard. 2012. The system of rice intensification: adapted practices, reported outcomes and their relevance in Cambodia. *Agricultural Systems* 113: 16–27.

Mal, P., A.R. Anik, S. Bauer, and P.M. Schmitz. 2012. Bt cotton adoption: A double-hurdle approach for North Indian farmers. *AgBioForum* 15 (3): 294–302.

Manda, J., A.D. Alene, C. Gardebroek, M. Kassie, and G. Tembo. 2015. Adoption and Impacts of Sustainable Agricultural Practices on Maize Yields and Incomes: Evidence from Rural Zambia. *Journal of Agricultural Economics*. DOI: 10.1111/1477-9552.12127.

Manjunatha and Kumar. 2015. Impact of National Food Security Mission (NFSM) on Input use, Production, Yield and Income in Karnataka. Institute for Social and Economic Change Bangalore.

Mansur, E.T., R. Mendelsohn, and W. Morrison. 2008. Climate change adaptation: A study of fuel choice and consumption in the US energy sector. *Journal of Environmental Economics and Management* 55: 175–193.

Marenya, P.P., and C.B. Barrett. 2007. Household-level determinants of adoption of improved natural resources management practices among smallholder farmers in western Kenya. *Food Policy* 32 (4): 515–536.

Marra, M., B. Hubbell, and G. Carlson. 2001. Information quality, technology depreciation, and Bt cotton adoption in the Southeast. *Journal of Agricultural and Resource Economics* 26 (1):158–175.

Mathenge, M.K., M. Smale, and D. Tschirley. 2015. Off-farm employment and input intensification among smallholder maize farmers in Kenya. *Journal of Agricultural Economics* 66 (2): 519–536.

Matuschke, I., and M. Qaim. 2009. The impact of social networks on hybrid seed adoption in India. *Agricultural Economics* 40 (5): 493–505.

Mazvimavi, K., and S. Twomlow. 2009. Socioeconomic and institutional factors influencing adoption of conservation farming by vulnerable households in Zimbabwe. *Agricultural systems* 101 (1): 20–29.

Mc Donald, A.J., et al. 2008. Stubborn facts: Still no evidence that system of rice intensification out-yields best management practices beyond Madagascar. *Field Crops Research* 108 (2008): 188–191.

Meshram, V., N. Chobitkar, V. Paigwar and S.R. Dhuware. 2012. Factors affecting on SRI system of Paddy Cultivation in Balaghat District of Madhya Pradesh. *Indian Research Journal of Extension Education* 1(Special Issue): 202–204.

Moser, C.M., and C.B. Barrett. 2003. The disappointing adoption dynamics of a yield-increasing, low external-input technology: The case of SRI in Madagascar. *Agricultural Systems* 76 (3): 1085–1100.

Namara, R.E., P. Weligamage, and R. Barker. 2003. *Prospects for adopting system of rice intensification in Sri Lanka: A socio economic assessment, research report, 75*. Colombo, Sri Lanka: International Water Management Institute.

Nirmala, K., R. Vasantha, and K. Supriya. 2015. Farmers knowledge on system of rice intensification (SRI) in Andhra Pradesh, India. *International Research Journal of Social Sciences* 4 (5): 5–11.

Noltze, M., S. Schwarze, and M. Qaim. 2012. Understanding the adoption of system technologies in smallholder agriculture: The system of rice intensification (SRI) in Timor Leste. *Agricultural systems* 108: 64–73.

Noltze, M., S. Schwarze, and M. Qaim. 2013. Impacts of natural resource management technologies on agricultural yield and household income: The system of rice intensification in Timor Leste. *Ecological Economics* 85: 59–68.

Ogada, M.J., G. Mwabu, and D. Muchai. 2014. Farm technology adoption in Kenya: A simultaneous estimation of inorganic fertilizer and improved maize variety adoption decisions. *Agricultural and Food Economics* 2 (1): 1–18.

Palanisami, K., C.R. Ranganathan, K.R. Kakumanu and Udaya N. Sekhar. 2011. A hybrid model to quantify the impact of climate change on agriculture in Godavari Basin, India. *Energy and Environmental Research* 1(1).

Palanisami, K., K.R. Karunakaran, U. Amarasinghe and C.R. Ranganathan. 2013. Doing different things or doing it differently? Rice intensification practices in 13 states of India. *Economic and Political Weekly* 18 (8): 51–58.

Pandey, S., D. Gauchan, M. Malabayabas, M. Bool-Emerick and B. Hardy. 2012. Patterns of adoption of improved rice varieties and farm-level impacts in stress-prone rainfed areas in South Asia. *The International Rice Research Institute (IRRI).*

Pandey, S., M.L. Velasco, and T. Yamano. 2015. Scientific strength in rice improvement programmes, varietal outputs and adoption of improved varieties in South Asia in Thomas S. Walker and Jeffrey Alwang, eds., in *Crop Improvement, Adoption, and Impacts of Improved Varieties in Food Crops in Sub Saharan Africa* 239–264. Boston, MA: CAB International.

Pathak, H., A.N. Tewari, S. Sankhyan, D.S. Dubey, U. Mina, V.K. Singh, N. Jain, and A. Bhatia. 2011. Direct seeded rice: Potential, performance and problems—a review. *Current Advance Agricultural Science* 3 (2): 77–88.

Pender, J., and B. Gebremedhin. 2008. Determinants of agricultural and land management practices and impacts on crop production and household income in the highlands of Tigray, Ethiopia. *Journal of African Economies* 17 (3): 395–450.

Prasad, C.S. 2007. Rethinking innovation and development: Insights from the system of rice intensification (SRI) in India. *The Innovation Journal*: *The Public Sector Innovation Journal* 12 (2): Article 3.

Prasad, C.S., D. Mohapatra, and P. Mishra. 2008. *Strengthening the learning alliance: Scaling up options for SRI in Orissa*. Bhubaneswar, Orissa: Xavier Institute of Management.

Rahm, M.R., and W.E. Huffman. 1984. The adoption of reduced tillage: the role of human capital and other variables. *American Journal of Agricultural Economics* 66 (4): 405–413.

Rajanna, M.P. 2010. Status paper on rice in Karnataka. *Rice Knowledge Management Portal (RKMP).* Hyderabad, India: Directorate of Rice Research http://rkmp.co.in/sites/default/files/ris/rice-state-wise/Status%20Paper%20on%20Rice,%2020.

Ramamoorthy, M. 2004. Studies on improved irrigation and weed management practices on growth and yield of squarely planted rice. M.Sc.(Ag.) Thesis, TNAU, Killikulam.

Reddy, V.R., P.P. Reddy, M.S. Reddy, and D.S. Rama Raju. 2005. Water use efficiency: A study of system of rice intensification (SRI) adoption in Andhra Pradesh. *Indian Journal of Agricultural Economics* 60 (3): 458–472.

Resurreccion, B. P., E.E. Sajor and H. Sophea. 2008. Gender dimensions of the adoption of the system of rice intensification (SRI) in Cambodia. *Asian Institute of Technology and Royal University of Phnom Penh.*

Revilla-Molina, I. M. et al. 2012. *Sowing the seeds of rice science: Achievements and future directions for training at IRRI.* International Rice Research Institute.

Rogers, E. 2003. *Diffusion of innovations*. New York: Free press.

Roodman, D. 2009. *Estimating fully observed recursive mixed-process models with cmp.* Available at SSRN 1392466.

Roodman, D. 2011. Fitting fully observed recursive mixed-process models with cmp. *The Stata Journal* 11 (2): 159–206.

Rosa Dias, P. 2010. Modelling opportunity in health under partial observability of circumstances. *Health Economics* 19: 252–264. DOI: 10.1002/hec.1584.

Ruppert, Peter, Elena Stancanelli, and Etienne Wasmer. 2009. Labor market outcomes: A transatlantic perspective.*Annals of Economics and Statistics* 95/96: 201–220.

SANDRP. 2014. Huge potential of system of rice intensification: SRI Needs government attention. Available at: http://sandrp.wordpress.com/2014/07/19/huge-potential-of-system-of-rice-intensification-sri-needs-more-attention/. Accessed on August 22, 2014.

Satyanarayana, A., T.M. Thiyagarajan, and N. Uphoff. 2007. Opportunities for water saving with higher yield from the system of rice intensification. *Irrigation Science* 25 (2): 99–115.

Senthilkumar, K., P.S. Bindraban, T.M. Thiyagarajan, N. De Ridder, and K.E. Giller. 2008. Modified rice cultivation in Tamil Nadu, India: Yield gains and farmers'(lack of) acceptance. *Agricultural Systems* 98 (2): 82–94.

Shiferaw, B.A., T.A. Kebede, and L. You. 2008. Technology adoption under seed access constraints and the economic impacts of improved pigeonpea varieties in Tanzania. *Agricultural Economics* 39 (3): 309–323.

Shiferaw, B., T. Kebede, M. Kassie, and M. Fisher. 2015. Market imperfections, access to information and technology adoption in Uganda: Challenges of overcoming multiple constraints. *Agricultural Economics* 46 (4): 475–488.

Sigdel, U.P., D. Devkota, N.R. Joshi, and N.R. Devkota. 2014. Impact of result demonstration on system of rice intensification (SRI) adoption among participant and non-participant farmers in Chitwan, Nepal. *International Journal of Agricultural Extension* 2 (3): 153–158.

Singh, V., and A. Kumar. 2014. Hybrid rice cultivation in Tarai region of Uttarakhand: An economic perspective of technology adoption. *International Journal of Basic and Applied Agricultural Research* 12 (2): 153–159.

Singh, R.K.P., K.M. Singh, and A. Kumar. 2015. A study on adoption of modern agricultural technologies at farm level in Bihar. *Economic Affairs* 60 (1): 49.

Sinha, S.K., and J. Talati. 2007. Productivity impacts of the System of Rice Intensification (SRI): A case study in West Bengal, India. *Agricultural Water Management* 87: 55–60.

Sita Devi, K., and T. Ponnarasi. 2009. An economic analysis of modern rice production technology and its adoption behaviour in Tamil Nadu. *Agricultural Economics Research Review* 22: 341–347.

Smale, M., P. Heisey, and H. Leathers. 1995. Maize of the ancestors and modern varieties: The microeconomics of high-yielding variety adoption in Malawi. *Economic Development and Cultural Change* 43 (2): 351–368.

SRI International Network and Resources Center. 2014. *SCI—the system of crop intensification: Agroecological innovations for improving agricultural production, food security, and resilience to climate change*. Cornell University, Ithaca, New York, and the Technical Centre for Agricultural and Rural Cooperation (CTA), Wageningen, Netherlands.

Stoop, W.A., N. Uphoff, and A. Kassam. 2002. A review of agricultural research issues raised by the system of rice intensification (SRI) from Madagascar: Opportunities for improving farming systems for resource-poor farmers. *Agricultural systems* 71 (3): 249–274.

Styger, E. 2014. *System of intensification research: A review-1993–2013*. In Presentation at the International Rice Congress, Bangkok, Thailand, October 31, 2014.

Styger and Jenkins. 2014. *Technical manual for SRI in West Africa: Improving and scaling up the system of rice intensification in West Africa*. Ithaca: SRI-Rice Centre, Cornell University. Available at: http://www.coraf.org/documents/publication/2014-11/sri_tech.pdf.

Takahashi, K. 2013. The roles of risk and ambiguity in the adoption of the system of rice intensification (SRI): Evidence from Indonesia. *Food security* 5 (4): 513–524.

Takahashi, K., and C.B. Barrett. 2013. The system of rice intensification and its impacts on household income and child schooling: evidence from rural Indonesia. *American Journal of Agricultural Economics* 96 (1): 269–289.

Teklewold, H., M. Kassie, and B. Shiferaw. 2013. Adoption of multiple sustainable agricultural practices in rural Ethiopia. *Journal of agricultural economics* 64 (3): 597–623.

Thakur, A.K and W.A. Stoop. 2015. The science behind the system of rice intensification: A climate-resilient technology. *Under Review for Publication in Indian Journal of plant physiology*.

Tiongco, M., and M. Hossain. 2016. Adoption of modern varieties and rice varietal diversity on household farms in Bangladesh (Vol. 22). *The International Food Policy Research Institute*.

Tsujimoto, Y., T. Horie, H. Randriamihary, T. Shiraiwa, and K. Homma. 2009. Soil management: The key factors for higher productivity in the fields utilizing the system of rice intensification (SRI) in the central highland of Madagascar. *Agricultural Systems* 100 (1): 61–71.

Uaiene, R. 2011. Determinants of agricultural technology adoption in Mozambique. In *African Crop Science Conference Proceedings* 10: 375–380.

Uphoff, N. 2002. Opportunities for raising yields by changing management practices: the system of rice intensification in Madagascar. In *Agroecological innovations: Increasing food production with participatory development*, ed., N. Uphoff, 145–161. London: Earthscan.

Uphoff, N. 2003. Higher yields with fewer external inputs? The system of rice intensification and potential contributions to agricultural sustainability. *International Journal of Agricultural Sustainability* 1 (1): 38–50.

Uphoff, N. 2007. Reducing the vulnerability of rural households through agroecological practice: Considering the system of rice intensification (SRI). *Mondes en Développement* 140 (4): 85–100.

Uphoff, N. 2010. *Agro-ecological approaches to climate proofing the agricultural sector while also raising its productivity in the 21st century*. Chapter to be published in proceedings from OECD Conference on soil systems and climate change, Madison, WI, June 2009.

Uphoff, N. 2012. *Raising smallholder food crop yields with climate-smart agricultural practices*. Report accompanying presentation on 'The System of Rice Intensification (SRI) and Beyond: Coping with Climate Change,' made at World Bank, Washington, DC.

Uphoff, N. 2016. *The system of rice intensification (SRI)-responses to frequently asked questions*. Available at: http://sri.cals.cornell.edu/aboutsri/SRI_FAQs_Uphoff_2016.pdf.

Uphoff, N., and A. Kassam. 2009. *Case study: System of rice intensification, in agricultural technologies for developing countries*, Final Report, Annex 3. European Parliament, Brussels.

Uphoff, N., A. Kassam, and R. Harwood. 2011. SRI as a methodology for raising crop and water productivity: Productive adaptations in rice agronomy and irrigation water management. *Paddy and Water Environment* 9 (1): 3–11.

V&A programme. 2009. Vulnerability and adaptation experiences from Rajasthan and Andhra Pradesh. In *The system of rice cultivation, SDC V&A programme, India. Approach to river basin management*. Earthscan, 166.

Vasishth, M. 2014. *Input use and agriculture in India: An analysis of system of rice intensification using I/O approach*. An Unpublished Master's Dissertation, TERI University, New Delhi.

Varma, P., and A Issar. 2016. Pricing to market behaviour of India's high value agri-food exporters: An empirical analysis of major destination markets. *Agricultural Economics* 47 (1): 129–137.

Veeraputhiran, R., R. Balasubramanian, and B.J. Pandian. 2014. Effect of mechanical weeding in system of rice intensification and its adoption. *Indian Journal of Weed Science* 46 (4): 383–385.

Vijayakumar, M., S. Ramesh, B. Chandrasekaran, and T.M. Thiyagarajan. 2006. Effect of system of rice intensification (SRI) practices on yield attributes, yield and water productivity of rice (*Oryza sativa L.*). *Research Journal of Agriculture and Biological Sciences* 2 (6): 236–242.

Wollni, M., D.R. Lee, and J.E. Thies. 2010. Conservation agriculture, organic marketing, and collective action in the Honduran hillsides. *Agricultural Economics* 41 (3–4): 373–384.

Wu, J.J., and B.A. Babcock. 1998. The choice of tillage, rotation, and soil testing practices: Economic and environmental implications. *American Journal of Agricultural Economics* 80: 494–511.

Wu, W., B. Ma, and N. Uphoff. 2015. A review of the system of rice intensification in China. *Plant and Soil* 393 (1–2): 361–381.

Yamano, T., M. Malabayabas, and M. Dar. 2013. Stress-tolerant rice in eastern India: Development and distribution. *STRASA, Economic Briefs, No-1. pp-3. IRRI, Bill and Melinda Gates Fondations, IRRI, Philipines*.

Yamano, T., A. Arouna, R.A. Labarta, Z.M. Huelgas, and S. Mohanty. 2016. Adoption and impacts of international rice research technologies. *Global Food Security* 8: 1–8.

Zhao, L., L. Wu, Y. Li, X. Lu, D. Zhu, and N. Uphoff. 2009. Influence of the system of rice intensification on rice yield and nitrogen and water use efficiency with different N application rates. *Experimental Agriculture* 45 (03): 275–286.

MIX
Papier aus verantwortungsvollen Quellen
Paper from responsible sources
FSC® C105338

If you have any concerns about our products,
you can contact us on
ProductSafety@springernature.com

In case Publisher is established outside the EU,
the EU authorized representative is:
Springer Nature Customer Service Center GmbH
Europaplatz 3, 69115 Heidelberg, Germany

Printed by Libri Plureos GmbH
in Hamburg, Germany